T0396526

A Brief Overview of China's ETS Pilots

Daiqing Zhao · Wenjun Wang
Zhigang Luo

A Brief Overview of China's ETS Pilots

Deconstruction and Assessment of Guangdong's Greenhouse Gas Emission Trading Mechanism

Daiqing Zhao
Guangzhou Institute of Energy Conversion
Chinese Academy of Sciences
Guangzhou, Guangdong, China

Zhigang Luo
Guangzhou Institute of Energy Conversion
Chinese Academy of Sciences
Guangzhou, Guangdong, China

Wenjun Wang
Guangzhou Institute of Energy Conversion
Chinese Academy of Sciences
Guangzhou, Guangdong, China

ISBN 978-981-13-1887-0 ISBN 978-981-13-1888-7 (eBook)
https://doi.org/10.1007/978-981-13-1888-7

Jointly published with China Environment Publishing Group Co., Ltd., Beijing, China

The print edition is not for sale in China Mainland. Customers from China Mainland please order the print book from: China Environment Publishing Group Co., Ltd.

Library of Congress Control Number: 2018950808

© China Environment Publishing Group Co., Ltd. and Springer Nature Singapore Pte Ltd. 2019
This work is subject to copyright. All rights are reserved by the Publishers, whether the whole or part of the material is concerned, specifically the rights of translation, reprinting, reuse of illustrations, recitation, broadcasting, reproduction on microfilms or in any other physical way, and transmission or information storage and retrieval, electronic adaptation, computer software, or by similar or dissimilar methodology now known or hereafter developed.
The use of general descriptive names, registered names, trademarks, service marks, etc. in this publication does not imply, even in the absence of a specific statement, that such names are exempt from the relevant protective laws and regulations and therefore free for general use.
The publishers, the authors and the editors are safe to assume that the advice and information in this book are believed to be true and accurate at the date of publication. Neither the publishers nor the authors or the editors give a warranty, express or implied, with respect to the material contained herein or for any errors or omissions that may have been made. The publishers remains neutral with regard to jurisdictional claims in published maps and institutional affiliations.

This Springer imprint is published by the registered company Springer Nature Singapore Pte Ltd.
The registered company address is: 152 Beach Road, #21-01/04 Gateway East, Singapore 189721, Singapore

Foreword

Carbon Emissions Trading Playing a Key Role in China's Ecological Construction

Guangdong Pilot ETS Offering Experiences for a Nationwide Carbon Market

Ecological construction is a strategic choice that tallies with the world development trends, while a green and low-carbon development pattern—a core content and top priority of ecological construction—has become a centerpiece that concerns the world sustainable development. Since the anthropogenic global warming has been intensively threatening the world ecological security and human survival, the different parts of the world have reached consensus to take concerted actions to handle the issue of climate change. In order to sustain socioeconomic prosperity while resolving the climatic challenge, all nations shall turn to a low-carbon and eco-friendly development pattern that harmonizes the relationship between mankind and nature, and ultimately transforms the human society from an industrial civilization to an ecological civilization. Under such circumstance, the environmental-bearing capacity seems to become an increasingly scarce resource, meaning that environmental capacity will become an indispensable production factor like labor force, capital, and land. In terms of the carbon Emissions Trading Scheme (ETS), it is a regime that treats emissions allowances as a scarce source and a production factor and exhibits their value by sales prices. Overall, the forging of carbon market is an essential part for China's ecological construction, because administration and transaction of emissions allowances will introduce revolutionary change upon the energy system, promote the transition of the social production and consumption patterns, and facilitates the popularization of a green and low-carbon socioeconomic growth pattern.

After the *Paris Agreement* was passed in 2015, all concerned nations, including China, were bearing an arduous task to cope with the pressing climatic issue. Being a nation of words and deeds, China has made great contributions in mitigating the global warming. From 2005 to 2017, China had lowered its CO_2 intensity of GDP by 45%, realizing in advance the reduction target of 40–45% in 2020 compared with 2005 that China pledged at the Copenhagen Climate Conference in 2009. With the *Paris Agreement*, China stated to cut its 2030 CO_2 intensity of GDP by 60–65% from the 2005 level, indicating an annual average decrease at above 4%, outpacing the average drop at 2% among the developed nations over 2005–2014. Moreover, China will make efforts to culminate its CO_2 emissions around 2030, and during this period, the annual drop in CO_2 intensity of GDP needs to reach 4–5% for the annual GDP growth rate would be 4–5%. In order to perform the obligations included in the *Paris Agreement*, China shall work even harder to make more contributions, which calls for a systematic backup from both institutions and policies. Therefore, China shall, on one hand, let the government play a leading role, hold onto the long-term low-carbon development strategy, carry out the near-term low-carbon development plan, insert restrictive emissions indicators into both the provincial and national 5-year plans, improve the fiscal and financial policies, constantly strengthen low-carbon technical norms and raise industry entry threshold. On the other hand, China shall give full play to the role of carbon market in achieving energy saving and emissions reduction. By combining with diverse policy instruments, China will be able to create an all-win landscape where there will be prosperous economy, improved environment, secure energy supply, and less CO_2 emissions.

Both the 18th National Congress of the Communist Party of China (CPC) and the Third Plenary Session of the 18th Central Committee of the CPC explicitly stated to forge ahead with the ecological civilization construction, let market play a decisive role in resource allocation, and actively carry out the pilot program about carbon emissions trading. China shall, through marketization and interest-driven mechanism, motivate individuals, enterprises, and governments to give their subjective initiative into full play, actively seek for low-carbon development, and alter the old growth pattern that overly relies on government plans and directives, so as to create a situation where all citizens take part in saving energy and cutting emissions. In October 2011, the National Development and Reform Commission (NDRC) released the *Notice on Carrying out the Work about the Carbon Emissions Trading Pilot Program in China* (NDRC Climate Change Dept [2011] No. 2601), which designates seven Chinese provinces and municipalities (incl. Guangdong Province) to take the lead in carrying out the ETS pilot program. Guangdong carbon market, which is the largest one among the seven pilot carbon markets, was officially launched on 19 December 2013. Through constant explorations and innovations, and based on steady progress, an open, transparent, well-organized, and efficiently operated Guangdong carbon market has basically taken shape to take charge for administration and trading of emissions allowances. During the 12th Five-Year plan period (2011–2015), Guangdong had cut the CO_2 intensity of GDP by 23.9%, exceeding the nationally restrictive target at 19.5%. By the end of May 2017, Guangdong carbon market had traded around 58.10 million tons of emissions allowances, holding 35.4% of the total of the 7 markets; earning total revenue at around 1.42 billion (bln) yuan, accounting for 36.9%. Thus, Guangdong carbon market was the first one of this type in China that broke the benchmark value at 1 bln yuan.

Guangdong is a fairly developed province in China, it is characterized by imbalanced regional economic growth, arduous task for cutting emissions, complete variety of industrial sectors and diversified emitters, which imply that Guangdong ETS design and operating experiences are worth of imitation and promotion, even its institutional layout and administrative accountability may inspire the building of a national carbon market. In a word, an all-round analysis of the seven pilot carbon markets, particularly Guangdong, will be of far-reaching significance for China's emissions reduction undertaking. At the time, when the pilot carbon markets are about to dock with the unified national carbon market, Guangzhou Institute of Energy Conversion (GIEC)—subsidiary to Chinese Academy of Sciences (CAS)—deliberately reviews and evaluates Guangdong ETS, decomposes the regime into several factors for revealing their designs, and for introducing the supporting policies behind the designing process. Filled with rich content, detailed cases and complete data, this book is able to transmit the ETS-related knowledge to the institutions, organizations, government officials, researchers, or corporate managers that are interested in carbon trading, or used as a textbook for training the talents in China carbon market.

He Jiankun

Beijing, China
July 2018

He Jiankun
Vice Chairman of China's National
Expert Panel on Climate Change,
Former Executive Vice President of
Tsinghua University

Preface

Addressing Climate change is today's common challenge in front of the mankind. In joining the global endeavor in mitigating climate risks, China helped concluding and enforcing the Paris Agreement, and delivered its Intended Nationally Determined Contributions (INDCs) to the UNFCCC[1] Secretariat in 2015, committing to form a national unified carbon Emissions Trading Scheme (ETS) steadily based on the pilots ETS programs, which is a crucial step in fulfilling its INDCs targets.

China launched the program of pilot ETS mechanism[2] in 2011, marking an official start of the nation's carbon market construction campaign. Guangdong Province—one of the two pilot provinces—opened its carbon market in 2013. To date, it has fulfilled three compliance periods with 100% of compliance rate for 2 years in a row, realizing smooth market performance and remarkable emissions cutbacks. As of 2016 end, Guangdong spot carbon market has traded 47.35 million tons (Mt) of emissions allowances, earning total turnover of 1261 million (mln) yuan, thus rising to China's largest and the world third largest carbon market. More than 70% of Guangdong-based covered enterprises lowered carbon intensity,[3] marking a prominent contribution in overfulfilling Guangdong's emissions reduction target, and in advancing industry transformation and upgrading during the 12th Five-Year-Plan period (2011–2015). Being China's first operated pilot carbon market at the provincial level, Guangdong has made several pioneering explorations by incorporating its characteristics, e.g., setting up a total allowances administration system under the emissions reduction target; managing the allowances to covered enterprises and new entrants in a separate manner; emissions from covered enterprises are under hierarchic (provincial/municipal level) administration; and integrating free allowances allocation with paid allocation. As a "pace setter, foregoer

[1] United Nations Framework Convention on Climate Change.

[2] In October 2011, the National Development and Reform Commission (NDRC) released the *Notice on Carrying out the Work of Carbon Emissions Trading Pilot Program in China* (No.2601 [2011]).

[3] "Carbon intensity" refers to carbon dioxide emissions per unit of GDP.

and tester," Guangdong ETS fully exhibits the characteristics of a provincial ETS in both framework and administration hierarchy.

In light of its carbon market construction agenda, a China-wide emissions trading scheme is about to be launched in 2017. However, unlike the seven pilot carbon markets, most of China's provinces and cities have little experience in this regard. They shall at first resolve several urgent questions before making such an attempt. For example, what are the interrelations between all elements under the ETS? What are the foremost questions under the framework of nationwide ETS? How to assess the ETS mechanism design? Guangzhou Institute of Energy Conversion, Chinese Academy of Sciences (GIEC, CAS)—is a leading think tank for building Guangdong ETS, sponsored by Guangdong ETS Pilot Program of China Clean Development Mechanism Fund (CDMFUND), Guangdong ETS Impact Assessment of UK Strategic Prosperity Fund (SPF), and Special Funds for Low-carbon Development of Guangdong Province. Under the guidance and elaborate organization of the NDRC and Guangdong Provincial Development and Reform Commission (GD DRC), GIEC joined in the formation of Guangdong ETS at the outset. It is deeply impressed that the ETS design is a fairly practical and systematic project. It involves multiple links as forming a management framework, defining covered enterprises, establishing emissions reduction targets and carbon offset rules, setting a cap on total allowances, determining allowances allocation methodologies, putting in place of a registration system, and guarding against market risks. Each link is separate but closely interacted. Therefore, a scientific analysis and assessment of each link and their effect is indispensable before an ETS is officially launched.

This book is an outgrowth of the joint efforts of GIEC's Energy Strategy Research Center and Non-carbon Energy Research Center. In this book, the authors share their thorough understanding of Guangdong ETS, exchanges with other pilot areas and new thoughts that were inspired by their peers. They break down the entire Guangdong ETS pilot program, dissect the macropolicies into the ideas for designing each link, and unfold the theoretical research process behind policy-making. Such an in-depth analysis will enlighten other provinces/cities that are interested in ETS, and promote smooth construction of a national uniform carbon market. We hereby send our gratitude to GD DRC's Department of Addressing Climate Change for their trust and support. Our thanks also go to other research institutions that have been working with us in building Guangdong ETS.

Guangzhou, China
December 2016

Daiqing Zhao

Contents

About the Authors

Daiqing Zhao research director and senior researcher of GIEC. She graduated from Tohoku University with a doctoral degree in Engineering. She is a member of the CAS "100 Talents Program", chief scientist of the "973 Program", expert member of Guangdong Government Decision-making Advisory Commission, environmental advisor of the Standing Committee of Guangdong Provincial People's Congress, Deputy Director of Combustion Institute of Chinese Society of Engineering Thermophysics, and part-time doctoral supervisor of University of Science and Technology of China and Nanjing University of Aeronautics and Astronautics.

Wenjun Wang Doctor of Economics and senior researcher of GIEC.

Zhigang Luo MBA and senior engineer of GIEC.

Xiaoling Qi Doctor of Engineering and senior researcher of GIEC.

Peng Wang Doctor of Engineering and senior researcher of GIEC.

Yuejun Luo Master of Science and assistant researcher of GIEC.

Pengcheng Xie Master of Management and engineer of GIEC.

Songyan Ren Master of Engineering and assistant researcher of GIEC.

Chubin Lin Doctor of Engineering and member of Postdoctoral Research Center of China Merchants Group.

Le Wang Doctoral student of CAS.

Guohui Gao Master of Economics and former researcher of Policy Studies Group of China (Guangzhou) Emissions Exchange.

Abbreviations

BAU	Business as Usual
BCRC	Beijing Climate Change Research Center
BP	British Petroleum
CAR	Climate Action Reserve
CARB	California Air Resources Board
CAS	Chinese Academy of Sciences
CCER	Chinese Certified Emission Reduction
CCR	Cost Containment Reserve
CCS	Carbon Capture and Storage
CDMFUND	China Clean Development Mechanism Fund
CEEX	China (Shenzhen) Emission Exchange
CER	Certification Emissions Reduction
CGE	Computable General Equilibrium
CNPC	China National Petroleum Corporation
CPC	Communist Party of China
CPI	Consumer Price Index
CR Power	China Resources Power Holdings Co., Ltd.
CSET	Chinese Society of Engineering Thermophysics
DEA	Data Envelopment Analysis
DID	Difference-in-Differences
DMU	Decision Making Unit
EEI	Energy Efficiency Index
EF	Energy Foundation
EIS	Electronic Information System
ETS	Emissions Trading Scheme
EU ETS	European Union Emissions Trading Scheme
GD LCPA	Guangdong Low-Carbon Economy Promotion Association
GDDRC	Guangdong Provincial Development and Reform Commission
GHG	Greenhouse gases
GIEC	Guangzhou Institute of Energy Conversion

GOF	Global Opportunities Fund
ICAP	International Carbon Action Partnership
IETA	International Emission Trading Association
IOU	Investor Owned Utilities
IPTS	Institute for Prospective Technological Studies
IRR	Internal Rate of Return
J-VER	Japan Verified Emissions Reduction
KETS	Korea Emissions Trading Scheme
KNN	k-Nearest Neighbor
KVAP	Keidanren Voluntary Emissions Action Plan
LSE	London School of Economics and Political Science
MGGRA	Midwestern Greenhouse Gas Reduction Accord
MRV	Monitoring, Reporting and Verification
NAP	National Allocation Plan
NAPCC	The National Action Plan on Climate Change
NDRC	National Development and Reform Commission
NZ ETS	New Zealand Emissions Trading Scheme
OTN	Obligation Transfer Numbers
PAT	Perform, Achieve and Trade
POLES	Prospective Outlook on Long-Term Energy Systems
POU	Publicly Owned Utilities
PPS	Production Possibility Set
PRIMES	Partial Equilibrium Model
REC	Renewable Energy Certificate
REDD	Reducing Emissions from Deforestation and Forest Degradation
RGGI	Regional Greenhouse Gas Initiative
ROI	Return on Investment
SAM	Social Accounting Matrix
SIC	Standard Industrial Classification
SPF	Strategic Prosperity Fund
TFP	Total Factor Productivity
UNFCCC	United Nations Framework Convention on Climate Change
VER	Voluntary Emission Reduction
WCI	Western Climate Action Initiative

Abstract

Deconstruction and Assessment of Guangdong Pilot Emissions Trading Scheme is a rich fruit of researchers' 5-year efforts in document compilation, surveys and studies, data analysis, consultations and discussions, and practical work in carrying out the pilot program. It covers all crucial factors that shall be considered for forming an ETS, ideas for designing each link, potential problems and difficulties as well as solutions. In addition, this book offers a quantitative assessment of Guangdong ETS from its operational efficiency, macro-and micro-influences, and draws some conclusions and inspirations that will benefit the formation of a nationwide ETS. Some relevant policies about Guangdong ETS are attached in the appendix for readers' reference.

This book consists of four parts: Part I introduces the global experiences in constructing carbon market, and analyzes the characteristics of Guangdong energy consumption and carbon emissions, in an aim to clarify the background for initiating ETS in the province. Part II elaborates on the formation and operation of Guangdong ETS and interprets the crucial elements during the pilot period (2011–2015), e.g., defining covered enterprises, calculating total allowances, developing allocation plans, designing the MRV[4] regime, and evaluating carbon market performance. In Part III, the authors use ICAP/CGE-GD and DEA models[5] to assess the macro- and micro-impact of Guangdong ETS, and find out that it lowers emissions reduction cost remarkably, but not efficient enough. Moreover, we also noticed that all elements involved in the ETS are closely interrelated, implying that when judging the input of an element is appropriate or not, we should take account of the input of other elements, instead of a cross-wise comparison of this single element or have it normalized. There is also an analysis of the factors that affect the ETS administration efficiency, including the regulated emissions quantity, total

[4]MRV: Monitoring, Reporting, and Verification.
[5]ICAP: International Carbon Action Partnership; CGE: Computable General Equilibrium; DEA: Data Envelopment Analysis.

amount of allowances, companies' profitability, and potentials in emissions cutbacks. Part IV (appendix) introduces the key policies that are unleashed while Guangdong ETS pilot program is implemented. They are listed in a chronological order for readers' convenience.

Chapter 1
Introduction

1.1 Introduction

In light of the directives of the 18th National Congress of the Communist Party of China (CPC) and the Third Plenary Session of the 18th Central Committee of the CPC, China shall vigorously press ahead with the construction of ecological civilization, let market play a decisive role in resource allocation, and actively carry out the pilot program for the carbon emissions trading scheme (ETS). In 2011, the National Development and Reform Commission (NDRC) released the *Notice on Carrying out the Work of Carbon Emissions Trading Pilot Program in China* (No. 2601 [2011]), which designates seven municipalities/provinces[1] to build an ETS that accords with China's special situations. After more than 2 years' of preparation, Guangdong ETS officially came online on December 19, 2013, and became the first one that had run through the entire operational process by July 15, 2014, including companies' emissions reporting, verification, allowances allocation and auction, and companies' compliance. As of December 1, 2015, China (Guangzhou) Emissions Exchange—an allowances bidding platform—had traded about 23.20 Mt of allowance—accounting for about 35.7% of the total trading volume of all pilot areas; and its accumulated revenue of 960 mln yuan held about 43.3% of the total revenue of all pilot areas, highlighting its position as the largest carbon market in China.

Guangdong carbon market has been so far operating smoothly. Local companies have learned more about the ETS and accepted this new environment governance mechanism. The covered enterprises are fairly observant, over 80% of them have achieved emissions cutbacks. Guangdong ETS has made several new breakthroughs, e.g., by following the principle of openness and transparency, it is the first one that made public the targets for setting a cap on total allowances; the aggregate allowances trading volume and revenue on the primary and secondary carbon

[1]The seven pilot areas are made up of the provinces of Guangdong and Hubei, and the municipalities of Beijing, Tianjin, Shanghai, Chongqing, and Shenzhen.

© China Environment Publishing Group Co., Ltd. and Springer Nature Singapore Pte Ltd. 2019
D. Zhao et al., *A Brief Overview of China's ETS Pilots*,
https://doi.org/10.1007/978-981-13-1888-7_1

markets rank the first place among all pilot areas; the entire province has become highly aware of the importance of emissions reduction; and more and more labor force has flocked into the low-carbon service sector. Besides, Guangdong took the lead in exercising allowances auction, drawing new entrants into emissions regulation, creating a carbon fund with the allowances trading revenue, and constantly innovating carbon finance products.

1.2 Formation Process of Guangdong ETS

Implementation of the ETS pilot program is a long-term endeavor. In light of the Guangdong Provincial Government Agenda, the ETS pilot program was carried out in three phases during the 12th Five-Year-Plan period (2011–2015):

Phase 1 (2012): Conduct preliminary researches, develop an implementation plan, and draw up the measures for allowances administration and transaction.
Phase 2 (2013): Put in place the ETS administration measures and implementation rules, develop an allowances allocation plan, build an Electronic Information System (EIS), and finally kick off the online transaction.
Phase 3 (2014–2015): Carry out the transaction activities in an all-round manner, assess the effect of relevant mechanisms, and plan for the work of emissions trading during the 13th Five-Year-Plan period (2016–2020).

1.2.1 Landmark Events

- In October 2011, the NDRC released the *Notice on Carrying out the Work of Carbon Emissions Trading Pilot Program in China* (No. 2601 [2011] NDRC Climate Change Department), which designates Guangdong and Hubei provinces and five municipalities to join in such pilot program.
- In May 2012, in order to absorb the ETS experiences both home and abroad, Guangdong competent departments, research institutions, and emissions exchanges, led by Guangdong Provincial Development and Reform Commission (GD DRC), visited UK for investigations and study, and then returned to China to exchange with Shanghai and other pilot areas.
- In August 2012, Zhuhai City of Guangdong hosted a seminar on the carbon emissions trading pilot program, which was attended by the leaders and experts from the NDRC Climate Change Department and other pilot areas. On this occasion, Guangdong introduced its ETS administration measures, the guidelines for companies' emissions verification and relevant provisions to draw comments and suggestions from other participants.

- In September 2012, at the opening ceremony of China (Guangzhou) Emissions Exchange, Guangdong Governor Zhu Xiaodan announced the official launching of Guangdong ETS pilot program. Xie Zhenhua, Deputy Director of NDRC; and Xu Shaohua, Executive Vice Governor of Guangdong, respectively, expressed their expectations for Guangdong ETS in their speeches.
- In June 2013, Xie Zhenhua, Deputy Director of NDRC, arrived at Shenzhen to preside over a special conference on the ETS pilot program joined by the representatives from all pilot areas. Xu Shaohua, Executive Vice Governor of Guangdong, pledged to kick off the emissions allowances trading by the end of the year.
- In July 2013, the Legal Affairs Office of Guangdong Government began to solicit public opinions for *Guangdong Carbon Emissions Allowances Administration and Transaction Measures*.
- In September 2013, the Guangdong Government held a special conference. Xu Shaohua, Executive Vice Governor, presided over a joint session on low-carbon emissions, where he listened to the report from GD DRC on the implementation progress of the Guangdong ETS pilot program.
- In November 2013, GD DRC released the *Notice on Carrying out the Work for Guangdong First Emissions Allowance Allocation* (for trial), and the name lists of reporting companies and covered enterprises.
- On December 16, 2013, Guangdong held an allowances auction—the first carbon market did so in China—fulfilling a trading volume of 3 Mt; yet the application volume on the same day was above 5.07 Mt, marking a shortfall in allowances supply.
- On December 19, 2013, Guangdong ETS officially came online, with the first-day trading volume hitting 120,029 tons and a total turnover of around 7.22 mln yuan. The strike price topped at 61 yuan/t and lowered at 60 yuan/t.
- In July 2014—deadline of the 2013 Compliance Period for Guangdong-based covered enterprises (the first ones under the ETS regulation), the companies' compliance rate hit high at 98.9%, and the allowances compliance rate was as even higher at 99.97%.
- In August 2014, GD DRC issued the *Allocation Plan of Guangdong's Carbon Emissions Allowances in 2014*, which makes public the name list of the covered enterprises and new entrants (including those expanded and rebuilt), and the allowances' calculation methodologies for 2014 Compliance Period.

1.2.2 Relevant Policies and Regulations

Since the ETS pilot program was kicked off, Guangdong Government has unleashed a slew of corresponding regulations and rules. As of December 30, 2015, a total of 11 policy papers were introduced to provide an institutional guarantee for the smooth operation of the ETS.

- On September 7, 2012, Guangdong Government released the *Notice on The Implementation Plan for Carrying out the Work for Guangdong Pilot Carbon Emission Trade Scheme* (No. 264 [2012] GD GOV), which presents the guidelines and objectives for the pilot program, and lays out overall work arrangement.
- On November 25, 2013, GD DRC issued the *Notice on Carrying out the Work for Guangdong First Emission Allowance Allocation* (For Trial) (No. 3537 [2013] GD DRC RES & ENV), which shows the list of the first batch of covered enterprises and new companies (including those expanded and rebuilt).
- On January 15, 2014, the 17th Session of the Standing Committee of the Twelfth Guangdong People's Congress adopted the *Temporary Measures for Guangdong Carbon Emission Management* (No. 197 GD GOV), which was put into effect since March 1, 2014.
- On February 28, 2014, GD DRC issued the Notice on Carrying out the Work for Emission Reporting and Verification of Companies in 2013 (No. 573 [2014] GD DRC RES & 1ENV), which made public the first batch of reporting companies and covered enterprises.
- On March 18, GD DRC issued the Notice on the Provisions for Emission Reporting and Verification of Companies (For Trial), the official Provisions for Emission Reporting and Verification of Companies (For Trial), and the Emission Reporting Guidelines for Companies (For Trial).
- On March 20, 2014, GD DRC issued the Guangdong Carbon Emission Allowance Management and Implementation Rules (For Trial), which clearly provides for allowance allocation, final settlement, and transaction.
- On June 9, 2014, GD DRC issued the *Notice on Urging covered enterprises to Pay off Their Emission Allowances* in 2013, which asked all these companies to register in Guangdong ETS and submit their allowance application and pay off the fees before July 15, 2014. GD DRC also released the Adjustment Program for Emission Allowance Allocation in 2013.
- On August 18, 2014, GD DRC issued the Notice on the Implementation Program for Guangdong Emission Allowance Allocation in 2014. In order to cope with changing economic situations and production fluctuations of enterprises, GD DRC decided to adjust 2014 allowance allocation by consulting with the assessment report of Allowance Allocation Assessment Panel, listening to the feedbacks from covered enterprises, and borrowing the experiences from both domestic and foreign carbon markets.
- On September 5, 2014, Guangdong Government made public the list of the first batch of recommended emission verification institutions, in light of the Temporary Measures for Guangdong Carbon Emission Management (No. 197 GD GOV), and through expert review and comprehensive assessment.
- On September 18, 2014, with the authorization of GD DRC, Guangzhou Carbon Emissions Exchange decided to launch the first allowance auction on September 26, 2014; the total allowances would be 2 Mt, in light of the Temporary Measures for Guangdong Carbon Emission Management (No. 197 GD GOV),

and Guangdong Carbon Emission Allowance Allocation Program in 2014 (No. 495 [2014] of the Climate Change Department of NDRC).
- On July 13, 2015, GD DRC issued the Notice on Guangdong Carbon Emission Allowance Allocation Program in 2015.

1.3 Experiences and Effects of Guangdong ETS

Guangdong is a leading developed province in China and also an epitome of the Chinese economic society, because of its unbalanced local economic growth, an arduous task for cutting carbon emissions and a complete industrial structure. In other words, the problems that Guangdong encounters in designing and operating an ETS shall be considered and resolved when building a nationwide ETS. Therefore, Guangdong will be a precursor for a nationwide ETS, with distinct values hardly outpaced by those pilot municipalities. Moreover, being one of the two provinces joining in the pilot program, Guangdong will set an example for other developed provinces. In short, a summarization of Guangdong's experiences in ETS design and operation is not only favorable for the province to refine management and trading of emission allowances, but also of far-reaching impact on forging ahead with a nationwide ETS.

(1) Building a clear-cut and forceful organizational structure, and constantly improving legislations.

Guangdong Provincial Committee of CPC and Government attach high attention to building an ETS. The low-carbon leading group, which is directly under the guidance of the Governor, maps out a general plan for carrying out the pilot program; while the Division of Climate Change of GD DRC takes charge of ETS design and operation. In order to smoothly carry out this initiative pilot program, GD DRC rallies intelligent sources from multiple academic sectors. It successively set up the ETS Research and Design Group, Allowance Management and Trading Workshop, and Allowance Allocation Assessment Panel. These entities contribute their ideas and thoughts in public policy management, jurisprudence, economics, environmental science, financial studies, statistics, and system software R&D. Thanks to these efforts, Guangdong has formed a working mechanism that enables all stakeholders, e.g., decision-making organs, research institutions, industry associations, and covered enterprises, to join in the formation of ETS, which has ensured a scientific, rational, and practical ETS design.

With regard to supporting laws and regulations, the Guangdong Government has come out with a slew of regulations and rules on the ETS, so as to create a solid legal basis for carrying out this pilot program. In January 2014, the Guangdong Government issued the Temporary Measures for Guangdong Carbon Emission Management (For Trial), which serves as a guideline for building the ETS. On this basis, GD DRC issued the Guangdong Carbon Emission Allowance Management and Implementation Rules (For Trial), and the Implementation Rules for Guangdong Corporate Emission Reporting and Verification of (For Trial), and released the emission reporting guidelines and verification specifications. Such regulations, together with the earlier rules on emission trading, have constituted an initial legal framework for ETS that is scientific, standardized, and carrying the characteristics of Guangdong.

(2) Maintaining exchanges and cooperation during the entire formation process of Guangdong ETS.

During the preliminary stage of ETS research and design, we were committed to learning the fundamental theories and global experiences. In addition to desk research, we visited EU, US, Japan, and Australia to learn from the front-line ETS experts and policy-makers. By following the global "cap-and-trade" principle, and considering the characteristics of Guangdong socio-economy and emission structure, we able to start the "top-down design" of Guangdong ETS.

While the ETS is being operated, we have been open to exchanges with the relevant research institutions, companies, and global experts, following the global carbon market trends, borrowing the strengths of foreign carbon markets to make up for our weaknesses and mitigate market risks. We have been closely watching the ETS operation so as to make the timely adjustment; interacting with other domestic carbon markets, and studying a scheme for linking Guangdong with other carbon markets. Our experts have maintained regular contacts with their counterparts in EU and UK, and held discussions with the experts of London School of Economics and Political Science (LSE) on effective assessment and policy adjustment.

Guangdong ETS is expected to be open and transparent from design to operation, and exhibit its global impact during interactions with the carbon markets both home and abroad. At the governmental level, the Guangdong Government conducted in-depth exchanges with the House of Commons Environmental Audit Committee and California Carbon Office on the implementation effects of policies. At the nongovernmental level, Guangdong Low-Carbon Economy Promotion Association (GD LCPA), joined by International Emission Trading Association (IETA), successively invited foreign low-carbon technology suppliers, emission reduction consulting firms, and allowance buyers to attend the exchange meetings with their partners in Guangdong. At the corporate level, several multinational enterprises, e.g., British Petroleum (BP) and London Office of China National Petroleum Corporation (CNPC), decided to join in Guangdong ETS, which has aroused the interest of the Chinese companies to devote themselves into the carbon market; KPMG and VERCO have made massive work in bringing forth carbon asset management strategy and enhancing low-carbon competitively of companies.

(3) Defining a scope of coverage in a scientific manner, and coordinating the relation between the cost and effect of emission reduction.

ETS construction involves seven components, i.e., the scope of coverage, targets for total emissions, allowance allocation rules, trading management, relevant laws,

MRV regulations, and links with other carbon markets. Among them, the scope of coverage is the fundamental component, since it determines the variety and quantity of covered enterprises and the total trading volume and allowance allocation approach, which will ultimately affect the ETS operational efficiency.

Guangdong ETS design group, in reference to the characteristics and requirements of carbon trading, and taking account of Guangdong carbon emission structure, economic plan, emission reduction targets, potentials for emission, cost for emission reduction, and data availability, built a "Selection and Assessment Model for Industrial Sectors" to define and assess the key industrial sectors to join in the ETS. Through calculation with this model, four industrial sectors of Guangdong were finally selected, i.e., electricity, cement, petrochemicals, and steel. In these sectors, there are 198 covered enterprises with annual carbon emissions exceeding 20,000 tons. These companies have large emissions, low-carbon leakage, accessible data, simple technologies (favorable for allowance allocation and verification), and great potentials for emission reduction. Their combined emissions account for around 50% of Guangdong's total emissions, indicating that their involvement in the ETS will help Guangdong fulfill its targets for emission reduction during the "12th Five-Year-Plan" Period. Moreover, thanks to less variety and quantity of covered enterprises, and lower expenses on technology and management, Guangdong ETS shows more advantages in operation efficiency, management, and technology costs than other pilot areas.

(4) Initiating a total allowance management system under the prerequisite for lowering carbon intensity, and exercising separate carbon budget management and accounting for old and new covered enterprises, which have resulted in remarkable emission reduction.

In light of China's overall GHG emission targets, Guangdong's development plan for key industries and target for controlling total energy consumption, Guangdong has conscientiously carried out the "cap-and-trade" mechanism by transforming the target for lowering carbon intensity into the upper limit of allowances, indicating that it has adopted the common systems in global carbon markets, and has the technical strength to link with other markets.

Guangdong is now at the stage of economic transition, so it is obliged to optimize the existing industrial sectors, foster growth of new productivity, and stabilize socioeconomic development. Based on the "cap-and-trade" mechanism, Guangdong initiated the separate emission accounting system for both old and new covered enterprises, established the basic principles for lowering the emissions of old companies, and strictly controlling the emissions of new companies, so as to ensure orderly management of provincial carbon emissions while sustaining economic growth.

Based on the "cap-and-trade" mechanism, the covered enterprises—holding about 60% of Guangdong's total emissions—have achieved significant results in cutting emissions by means of technical transformation and building internal energy management system. In 2013, the old covered enterprises in electric and petrochemical sectors lowered absolute emissions, i.e., their emissions fell 5–7% from 2012. All covered enterprises saw their emissions of per unit production drop by 2–10%, marking a great contribution in fulfilling Guangdong's target for lowering carbon intensity.

(5) Lowering social cost for emission reduction and helping Guangdong transit to a low-carbon economy.

A multi-tier carbon market system has initially taken shape in Guangdong. It covers an allowance trading market and a verified VER market that both serve productive enterprises, a Carbon Inclusive System that benefits residents, as well as a linking mechanism between markets. Under such a multi-tier framework, covered enterprises are able to purchase allowances or a certain proportion of verified VER volume to offset their emissions, which helps covered enterprises lower cost for cutting emissions, and stimulates noncompliance companies to join in emission reduction. Under the Carbon Inclusive System, the individuals that have extravagant carbon consumption shall bear some cost for environmental governance, while those that insist on green consumption and environmental protection will be rewarded.

In retrospect to the years' socioeconomic developments of Guangdong, GIEC built a CGE Model that reflects the dynamic characteristics of the industrial structure and energy consumption structure of 33 sectors. They used this model to simulate Guangdong energy consumption and carbon emissions by 2020, and quantitatively assess the implications of ETS on the provincial socio-economy, cost of emission reduction, carbon price, and employment, and has come to the following conclusions: First, ETS is able to cut Guangdong total CO_2 emissions. By 2020, Guangdong emissions are predicted to reach 690 Mt, lower by 30–60 Mt than base scenario; it will be enough to fulfill the target for lowering carbon intensity by 45% at that time, as long as GDP growth remains at 7.5% (per capita GDP and carbon emissions in 2020 will increase four times and two times, respectively, from 2005). Second, owing to the cap on total emissions and the emission constraints on industrial sectors, Guangdong GDP growth may drop 1.23–1.58% by 2020. However, if the four sectors could join in the ETS, the lower cost for emission reduction will mitigate the GDP loss by 0.10%, i.e., to save the economic loss of 9 billion yuan, which will, in turn, greatly improve social welfare system. Third, the cap on emissions somewhat impacts the employment in the energy-intensive sectors, but creates new job opportunities on the whole, especially in the tertiary industry. The government shall, by means of increasing vocational education and training, help divert the labor force into the tertiary industry and service sector, which will mitigate the adverse impact on manufacturing and social stability.

The analysis of the macroeconomic implications of ETS with the CGE Model is from the perspective of the balanced economic system, and on account of the assumptions of full market competition, equilibrium carbon price based on market transparency, and full exploitation of reduction potentials. However, these assumptions may somewhat vary from the actual situations in Guangdong carbon market, indicating that the simulation results will have some ambiguities.

(6) Playing a significant role in expediting the elimination of backward production capacity and promoting industry transition and upgrading.

In addition to achieving cost-effective emission reduction, ETS shall be able to promote and guide the fulfillment of socioeconomic targets. This flexible carbon trading mechanism shall integrate industry transition and upgrading with emission reduction, in an aim to gradually raise Guangdong's energy-use efficiency, lower its total emissions, and upgrade its industrial structure and energy consumption structure. Guangdong ETS adopts different allowance allocation approaches for different industries. A set of strict allocation criteria encourages the companies with high emissions of per unit production to turn to high-tech and low-emission technologies and energy-saving equipment. As for the enterprises that are blacklisted by national and provincial departments for their outdated capacity, they have been pacing up the voluntary elimination of backward capacity or industry transferring driven by the ETS constraints and profits from allowance trading, which will quicken the readjustment of Guangdong industrial layout, and balance regional economic development.

(7) Bolstering the development of low-carbon service sector and increasing job opportunities in this regard, greatly enhancing the ability and awareness of companies for controlling emissions.

After the ETS was launched, Guangdong has seen increasing demand for carbon-related organizational management, operational maintenance, third-party service, trading finance and publicity, which has given rise to a low-carbon service sector constituted by emission consulting firms, carbon asset management institutions, and exchange intermediaries. In 2013, Guangdong established 16 institutions for carbon emission verification, which hire almost 100 inspectors. If the consulting and service staff members in low-carbon service sector were counted in, then Guangdong ETS has directly created about 20,000 jobs. Besides, an analysis based on CGE Model shows that Guangdong carbon market will increase additional 40,000–50,000 jobs by 2020. Along with the continuous development of ETS, carbon trading will play a more significant role in industry transition and job creation.

Upon launching the ETS, how to use emission allowances efficiently has become one of the factors that affect corporate decision-making. Currently, the companies that have poured funds for technical upgrading and transformation are incorporating the profit from emission reduction into their return on investment (ROI). Several Chinese major groups, e.g., Guangdong Yudean Group and China Resources Power Holdings Co., Ltd. ("CR Power"), attach great attention to carbon budget and emission management, and they have come up with the group-wide carbon asset management rules.

(8) Guangdong ETS is receiving attention from domestic and overseas carbon markets.

Guangdong carbon market features large capacity, extensive coverage, and diverse allocation patterns. It is the first one to create an allowance auction in China. Such characteristics have made Guangdong one of the most influential carbon markets both home and abroad. The international media have been closely watching the Guangdong carbon market. Reuters complimented Guangdong competent departments as "brave in transition and daring to be the first" for its boldness in borrowing experiences from mature carbon markets, initiating allowance auction in China, motivating companies to pay for emission allowances, and adding new companies into allowance management (their operation relies on allocation purchase). Guangdong carbon market has risen to the world's second largest emission allowance market. In addition, Point Carbon spoke highly of Guangdong ETS for its transparency and data publicity, believing that it is a correct step toward building an open, just, and fair carbon market. IETA also expressed interest in Guangdong carbon market, it co-sponsored two seminars in Guangdong with GD LCPA, in an aim to encourage companies to learn from mature carbon markets in carbon asset management, and help Guangdong-based covered enterprises to actively deal with carbon trading.

In addition, the EU Commission, the UK Parliament and California Air Resources Board (CARB) also maintain regular policy exchanges and dialogue with Guangdong to discuss on the possibility of cooperation. The US Energy Foundation (EF), the UK Global Opportunities Fund (GOF) and other international funds show expectation for Guangdong carbon market; they have covered Guangdong ETS into one of their key funding projects to support Guangdong research institutions and their global counterparts to work together in frontier research about carbon trading.

Chapter 2
Global ETS Operation and Their Merits and Demerits

Since the Kyoto Protocol (KP) was signed in 1997, all parties to the KP have been actively exploring the path to transit to a low-carbon economy, and using market mechanism to cut Greenhouse Gas (GHG) emissions and save the cost for emission reduction. On January 1, 2005, European Union Emission Trading Scheme (EU ETS) was launched, which was followed by Regional Greenhouse Gas Initiative (RGGI), Midwestern Greenhouse Gas Reduction Accord (MGGRA), Western Climate Action Initiative (WCI), and California Cap-and-Trade. Australia achieved significant reduction results from the fixed-carbon emission reduction system. New Zealand Emission Trading Scheme (NZ ETS) has included agriculture, fishery, and forestry into governance, since agriculture is the country's pillar industry. Japan started VER since 1997 and has formed a cap-and-trade system at the municipal level.

In addition to the contracting parties to the UNFCCC (see Annex I), some non-contracting parties also pledged to join in this emission reduction campaign. For example, China defined seven provinces and municipalities to carry out the ETS pilot program, and then announced at the Paris Climate Conference in 2015 (COP21) to launch the nationwide carbon market in around 2017. Mexico set the targets and specific measures for emission reduction by 2020. South Korea started exercising the ETS since 2015. India has been making efforts in developing renewables and a market-oriented energy mechanism.

2.1 Construction and Operation of Global ETS

2.1.1 EU ETS

EU has all along been an initiator and forerunner in responding to climate change. From 1992 to 2008, the UK had promulgated 73 policies to deal with the challenges posed by climate change, and has achieved significant results [1]. After the KP took

© China Environment Publishing Group Co., Ltd. and Springer Nature Singapore Pte Ltd. 2019
D. Zhao et al., *A Brief Overview of China's ETS Pilots*,
https://doi.org/10.1007/978-981-13-1888-7_2

effective since 2005, EU launched the ETS—the largest GHG emission trading system in the world, for the purpose of helping member countries lower emission reduction cost. To date, EU ETS has been in operation for almost 10 years, and passed through the preliminary phase (2005–2007) of "learning-by-doing" and the interim phase (2008–2012) that is filled with drastic market fluctuations; it is now at the third phase (2013–2020) which is considered as a "post-KP period" that features adjustment of both market and institutions. The development of EU ETS is a process of sparse allowance management transfer to unified management and also a gradual improvement of the carbon market.

EU ETS, a strategic policy to cope with climate change, covered 50% of EU CO_2 emissions when launched. It scoped over 11,500 fixed emission installations from electricity generators, heat and steam production, mineral oil refineries, coke ovens, ferrous metals production and processing, cement, lime glass, bricks and ceramics, pulp and paper of 28 member countries. Six types of GHGs were emitted from these sectors (CO_2, CH_4, N_2O, HFCs, PFCs, and SF_6) were covered [2]. From 2010, more than 4000 aviation operators were also scoped into EU ETS. The scope of EU ETS in Phase 3 has further extended to chemical, synthetic ammonia, nonferrous melting, and aluminum sectors. PFCs arising from electrolytic aluminum, and N_2O arising from chemistry, ammonia, aluminum, nitric acid, adipic acid, and glyoxylic acid were covered.

(1) Emission reduction targets

EU ETS set three phases to reach its emission targets. Phase 1 (2005–2007): to fulfill 45% of EU commitment target under KP. Phase 2 (2008–2012): each EU member state cuts average 6.5% emissions based on 2005 level. The total target of these two phases is in accord with the target of KP's first commitment period (2008–2012), which is by 2012, the total emissions of 15 European countries will decrease 8% compared to the emission level of 1990. In Phase 3 (2012–2020), EU emission reduction target will be formulated according to European "20/20/20" targets[1]; i.e., by 2020, European GHG emissions will be cut 20% based on 1990 level. In 2014, The European Council approved a more stringent emission reduction target for 2030, that: abatement of GHG emissions shall reach to at least 40% from 1990 level. In light of the overall emission reduction targets, EU ETS allowances allocated in the three phases will decrease gradually, i.e., Phase 1 preserves 2299 Mt CO_{2e}/a; Phase 2 preserves 2081 Mt CO_{2e}/a; Phase 3 cuts preserved allowances by 1.74% annually; and Phase 4 cuts preserved allowances by 2.2% annually.

[1]By 2020, European GHG emissions will decrease by 20% based on 1990 level; the share of renewables in European total energy consumption will to 20%; European energy-use efficiency will rise by 20%.

(2) Allowance allocation scheme

The allowance allocation of EU ETS in Phases 1 and 2 was based on National Allocation Plan (NAP). European Commission allocated allowances to each member country based on their reported historical CO_2 emissions, such allocation approach is called "Grandfathering". In Phase 1, 95% of the allowances were allocated for free, and each member country was allowed to purchase no more than 5% of allowances through auction. In Phase 3, European Commission canceled NAP, but adopted "Benchmarking", which means there was a benchmark for allocation based on CO_2 emissions per unit of production in different industrial sectors and production activities; the proportion of free allocation was gradually lowered, while the proportion of allocation auction would be increased.

(3) Flexible mechanisms

The surplus allowances in Phase 1 could be banked or Phase 2, with an aim to encourage emitters to cut more emissions according to their actual emission volume and allowance price, and maintain vigor and continuity of the secondary allowance market.

(4) Compliance

EU ETS imposes penalty on non-compliance companies. In Phase 1, one ton of excessive emission will be fined €40/CO_{2e}. In Phase 2, one ton of excessive emission will be fined €100/CO_{2e}. In Phase 3, the amount of fine will increase along with the European Consumer Price Index (CPI). The compliance ratio in 2005–2009 was above 98% and reached 100% in 2006–2008.

(5) Carbon market performance

EU carbon market has been developing rapidly. In 2010, the trading volumes in the EU carbon market accounted for 84% of the global total trading. EU carbon market is the largest market of this type to date. By 2012, the allowances traded in the EU carbon market reached 7.9 billion CO_{2e}, which was 17 times more than that in 2006. See Fig. 2.1 [3].

(6) Effectiveness assessment

In Phases 1 and 2, EU allocated all 3 years' allowances for free. Allowances were based on historical emissions of member countries without considering the impact of economic fluctuations, and lack of allowance assessment and an adjustment mechanism. In 2008 when a global financial crisis broke out, the European industrial activities fell into a downturn, decreasing carbon emissions resulted in massive surplus allowances. Moreover, a large quantity of low-price carbon offset credits was used for commitment, which exacerbated allowance excess, and pulled down allowance prices further. At the end of Phase 2, the total allowances surplus of European companies were about 2100 Mt CO_{2e}, which equaled to an annual allowance volume of EU ETS, thus causing the allowance price to plunge again after a 2-year stable implement.

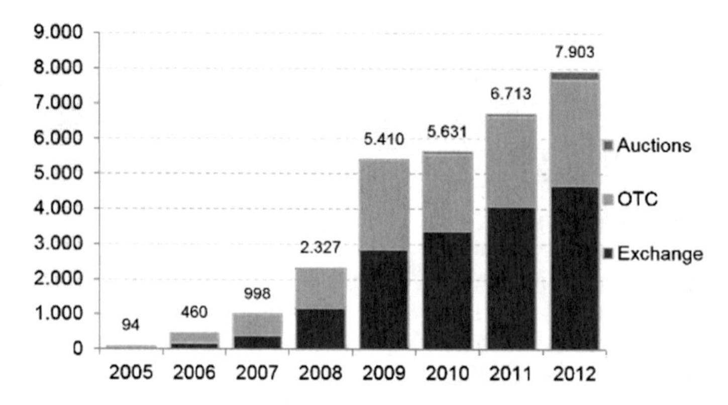

Fig. 2.1 EU ETS Phase 1 and Phase 2 Allowance Trading Volume (bln CO_2). *Source* "Evolution of the European carbon market", EU Emission Trading System, EU Action, Climate Action, European Commission official website. https://ec.europa.eu/clima/policies/ets/pre2013_en

In face of massive surplus allowances and gloomy market, EU was forced to reform the allowance allocation in Phase 3. First, EU adopted a unified allowance calculation to prevent unequal allocation among member countries. In order to reduce allowance inventory and cancel the "windfall profit" of the electricity sector, all of the allowances allocated to the electricity sector was subject to auction. In contrast, the industrial sectors facing international competition were receiving free allowances to safeguard their competitiveness and to prevent carbon leakage. For example, 85% of the allowances given to aviation operators were free of charge. However, the ratio of free allowances will decrease year on year, i.e., from 80% in 2013 to 30% in 2020 [4]. Second, EU postponed allowance auction. The 900 Mt CO_{2e} of allowances in Phase 3 are preserved for auction in 2019–2020, which, in turn, cuts the allowance auction in 2014, 2015, and 2016 by 400, 300, and 200 Mt, respectively. Consequently, the ratio of allowance auction has been increasing after entering Phase 3: exceeding 50% in 2013, reaching 70% in 2020, and 100% in 2027 [5]. Third, only the emissions based on Certification Emissions Reduction (CER)—from either the least developed countries or any country that has signed a bilateral agreement with EU—are accepted for compliance [6]. Last, EU plans to introduce an allowance reserve mechanism since 2018 for market stable to resolve excessive allowance allocation in the long run. Meanwhile, while considering the allowance supply and demand, EU will adjust the volume of allowance auction to regulate the supply–demand pattern, and strengthen the capacity of EU ETS to resist market impact [7].

(7) Carbon Market Linkage

A global common mechanism for tackling climate change through negotiations is a long-term task, but EU has been attempting to link and cooperate with worldwide carbon markets, and help different countries establish the "cap-and-trade" system. EU believes that the market linkage shall meet the following conditions:

- Compatibility. Different trading systems shall have the same operational environment, e.g., 1 ton of CO_2 shall have the same equivalent value within different systems;
- Equivalent policy imperative;
- There is a cap on total emissions for each ETS scheme.

In light of the above requirements, EU and Australia used to negotiate on linking their ETS, but failed to reach any consensus, as Australia repealed its ETS in 2014 [8].

In order to encourage the development of global carbon market, at the time that the Paris Climate Conference was about to reach a new global climate change agreement, EU expects to define the rules for the international carbon market. EU holds that a new global carbon market mechanism shall be built, which is similar to the KP-based Clean Development Mechanism (CDM) and Joint Implementation (JI), in an aim to deepen international collaboration in emission reduction.

2.1.2 ETS in North America

Before the book was finalized, the US has not yet developed a nationwide cap-and-trade program, but several states have made their first move. Certain regional cap-and-trade programs, particularly the Regional Greenhouse Gas Initiative, Midwest Greenhouse Gas Reduction Accord, and Western Climate Initiative and California Air Resource Board, took shape one after another.

(1) Regional Greenhouse Gas Initiative

The Regional Greenhouse Gas Initiative (RGGI), which was officially launched in 2009, is the first mandatory market-based program in the US to reduce GHG emissions. The RGGI is a cooperative effort of nine Northeast and Mid-Atlantic US states—Connecticut, Delaware, Maine, Maryland, Massachusetts, New Hampshire, New York, Rhode Island, and Vermont—to regulate and reduce CO_2 emissions from 225 power plants (in operation of 500–600 generator units). For the first 3-year period (2009–2011), the cap for 10 states was 188 Mt. For the second 3-year period (2012–2014), the cap for the 9 RGGI states (New Jersey dropped out) was 165 Mt per year in 2012 and 2013. A self-assessment in early 2014 revealed that the reserved allowances were far more excessive than the actual emissions. Based on this result, several amendments were made to the RGGI Model Rule: the Emissions Cap in 2014 will be within 91 Mt. The Model Rule language maintains the original 2.5% per year reduction to the regional RGGI cap for the years 2015 through 2020 [9]. In the meantime, the RGGI will, based on the surplus allowances in the prior period, preserve a portion of allowances for macro regulation [10].

The RGGI was the world's first cap-and-trade program that allocated all emissions allowances through auctions. Regional auctions are held on a quarterly basis. They were initially conducted in a single round using sealed-bid or uniform price format, and later conducted in multiple rounds using ascending price formats. The unsold allowances in each auction will be transferred to the next auction. When entering into a new period, the RGGI will evaluate the current allowances' supply and demand, actual emissions and compliance performance of covered enterprises, and then decide to either revoke the undistributed allowances or transfer them for auction in the coming period. The first upset price in the 2009 auctions was set as $1.86/t, which was adjusted up in each auction after 2009 in reference to the CPI fluctuations. In addition to the upset price, the RGGI will bring in more allowances and price regulation mechanisms for easing fluctuations of allowances market. In light of the amended RGGI Model Rule, the unsold allowances left from 2012 and 2013 auctions will not be transferred into new auctions since 2014, and the Cost Containment Reserve (CCR) was introduced. The CCR would consist of a fixed quantity of allowances, in addition to the cap, that would be held in reserve, and are only to be made available for sale if allowance prices were to exceed predefined price levels.

In terms of complementary mechanisms, the RGGI covered enterprises, in addition to obtaining allowances through auctions, may purchase offset credits to deduct their emissions. Offset credits in the following five project categories may be eligible for use in the RGGI State regulations: (i) Landfill methane capture and destruction; (ii) Reduction in emissions of sulfur hexafluoride (SF_6) in the electric power sector; (iii) Sequestration of carbon due to the U.S. forest projects (reforestation, improved forest management, and avoided conversion) or afforestation (for CT and NY only); (iv) Reduction or avoidance of CO_2 emissions from natural gas, oil, or propane end-use combustion due to end-use energy efficiency in the building sector; (v) Avoided methane emissions from agricultural manure management operations.

All offset projects must be located within one of the RGGI States, or in any other state that agrees to implement the RGGI emissions reduction criteria; and commencement of these offset projects should be restricted in a certain period. Finally, the deduction proportion of CO_2 offset credits shall tie to the allowance auction prices [11].

(2) Midwest Greenhouse Gas Reduction Accord

The Midwest Greenhouse Gas Reduction Accord (MGGRA) is a regional agreement by six governors of states in the US Midwest (Minnesota, Wisconsin, Illinois, Iowa, Michigan, and Kansas) and the Canadian Province of Manitoba. It covered electricity generation and imports, industrial process sources, transportation fuels, and commercial sectors. The Accord envisions cutting GHG emissions 20% below 2005 levels by 2020, and 80% below 2005 levels by 2050.

The Accord has been inactive to date with allowance allocation not started, yet a special Advisory Group—constituted by the environmental division, industrial department and administrative department in all of the signatories to the Accord—has begun to provide recommendations regarding the implementation of the Accord. Allowance allocation is to be calculated in a uniform approach, granting the reward of allowances to the early starters, as well as taking account of emissions increase resulting from economic and population growth in any regulated state and province. Allowance allocation is first subject to an auction and free distribution, and gradually replaced by complete auction. The Accord allows for offset credits, which are no more than 20% of the allowance cap. The Accord shall be developed in a manner that facilitates linkage with other programs like RGGI, WCI, and EU ETS.

(3) Western Climate Initiative

The Western Climate Initiative (WCI) is a cooperative effort of seven U.S. states (Oregon, California, Washington, New Mexico, Arizona, Montana, and Utah) and four Canadian provinces (British Columbia, Manitoba, Ontario, and Quebec). The states of Alaska, Colorado, Idaho, Kansas, Nevada, and Wyoming participate as observers, as do the Canadian province of Saskatchewan and the Mexican border states of Baja California, Chihuahua, Coahuila, Nuevo Leon, Sonora, and Tamaulipas.

Western Climate Initiative funded a nonprofit corporation in 2011 to provide administrative and technical services to support the implementation of state and provincial GHG trading programs.

Beginning in 2012, the initial compliance period , the program will cover 90% emissions of involved states with an overall emissions reduction objective of lowering 2020 emissions by 15% from 2005 levels. The scoped sectors include electricity, including electricity imports; fossil fuel combustion at large sources; and industrial process emissions. The second compliance period would begin in 2015, the program would expand to cover the combustion of natural gas and diesel oil at transportation sector, fossil fuels used for residents, and commerce, as well as other industrial fuels.

Generally, allowance distribution will be done independently by each WCI Partner jurisdiction. For the first compliance period, the WCI Partner jurisdictions will auction a minimum of 10% of the allowance budget, and to increase the minimum percentage to reach 25% in 2020.

In order to encourage early initiators prior to the start of the program, WCI considering to issue "Early Reduction Allowances" as a reward to provide incentives for emission reduction. Each Partner will have the discretion to bank allowances to the next phase.

The WCI Partner jurisdiction will limit the use of all offset credits. The proportion would be no more than 49% of the total cap during 2012–2020. Each WCI Partner jurisdiction will have the discretion to set more stringent rules for abatement.

Through a review of the WCI operation, it is only California and Quebec that have implemented the cap-and-trade mechanism (with the former becoming a separate executor with the "California Air Resource Board"), and jointly organized allowance auctions. Other states and provinces are yet to publish any substantial progress in cutting emissions.

(4) California Air Resource Board

California Air Resource Board (CARB), which was created in 1967 to aggressively address the serious issue of air pollution in the state, saw its role expanded by the Global Warming Solutions Act of 2006 (Assembly Bill 32 or "AB 32") to development and oversight of California's main GHG reduction programs.

The CARB regulates 85% of California GHG emissions generated from about 600 emitters with annual emissions above 25,000 tons. They are industrial companies; electric power generators, transmitters, and distributors; fuel producers, suppliers, and importers; residential and commercial natural gas distributors; LPG producers; producers and suppliers of transportation fuels and biomass fuels. The CARB regulation also extends to the six KP-restricted GHGs (CO_2, CH_4, N_2O, HFC_S, PFC_S, and SF6) and NF_3.

California has set the target to roll back carbon emissions to 1990 levels by 2020, marking an emission drop of around 9% from 2005 levels.

Transportation accounts for 37% of GHG emissions in California. Reducing GHG emissions from this source category is vital in achieving the goals of AB 32. Understanding this challenge, the state government exercises emission regulation over the fuel stations that import fuels to California, i.e., to exert pressure upon the supply end [12]. Besides, with 31% of the state electricity demand satisfied by outbound sources, the CARB has to impose more cost on electricity importers to curtail GHG emissions.

The CARB involves three phases: in Phase 1 (2013), the total allowances were 98% of the emissions of the covered enterprises in 2012. In Phase 2 (2014), the allowances dropped 3% from 2013 levels. In Phase 3 (2015–2020), the allowances will drop 3% year on year from 2013 levels.

The CARB allowance auction is based on a quarterly basis. Free allowances are granted to the industrial installations that are prone to industry transfer under the cost pressure from the cap-and-trade mechanism (a latent danger for carbon leakage). The amount of free allowances is calculated on basis of carbon intensity baseline, so the companies with lower carbon intensity will have access to more free allowances. The baseline is constantly updated in light of the carbon intensity reported by companies, so as to encourage the low-carbon development of California-based companies [12]. In contrast, the free allowances granted to electricity producers and distributors are merely 24% of their total demand for allowances in 2013–2020. For other industries, the percentage of free allowances range from 30 to 100% in light of their carbon leakage. In the meantime, in order to guard against drastic price fluctuations in allowance auctions, California sets the upset price at \$10 t/CO_2, and reserves 5% of the total allowances for market regulation, i.e., setting the price for reserved allowances is a means to control price ceiling in auctions.

California allows offset credits of nonlocal projects to offset the local emissions. Companies may use the offset credits to meet up to 8% of their compliance obligations. Yet there are also strict rules on the operation time and location of the offset projects. The Climate Action Reserve (CAR) only approves the offset credits from four categories of projects that are about forestation, urban forestry, livestock breeding, and ozone depletion. The offset credits from the projects in developing countries may not be above 25% of the total allowances in 2013; and such percentage may not be above 50% through 2014–2020. California also accepts the offsets from the Reducing Emissions from Deforestation and Forest Degradation (REDD) in developing countries. The offset credits approval based on the CAR methodology is about 8.70 Mt, which is estimated at 125 Mt in 2013–2020, i.e., equal to about 4% of the total emissions during this period.

In terms of the allowance allocation approaches, several US carbon emission trading mechanisms adopt allowance auction, which takes account of the emissions reduction cost and actual allowance demand of covered enterprises, and keeps from excessive allocation. However, the RGGI operation experience shows the Phase 2 allowance allocation was excessive, because of too many allowances placed in the auction and extensively sold at lower prices. Such experience tells that allowance allocation, even by means of an auction, shall be adjusted amid economic fluctuations and abiding by strict allocation principles, instead of setting any unrealistic lower upset prices.

The US initiated SO_2 and NOx emissions trading as early as in the 1990s, the mature experiences therefrom have laid a solid foundation for each state to develop their own cap-and-trade system. Local governments have been actively playing a lead role in caring out carbon emissions reduction programs and mapping out trading rules. Although a nationwide GHG emissions reduction campaign is still in absence in the US, the regional emissions reduction endeavors have formed a bottom-up driving force and good example, which will stimulate the promulgation of a nationwide emissions reduction policy. In contrast to a nationwide carbon market, regional cap-and-trade mechanisms—though help to fulfill regional emissions reduction targets—have certain limitations since they are separate from each other, the varied allowance prices and different marginal reduction cost may hold back effective allowance allocation and lower reduction efficiency [13].

2.1.3 ETS in Australia

Australia carbon emission trading is carried out in several phases, and each phase takes on different characteristics. Phase 1 (July 2012–July 2015) was actually a phase featuring fixed pricing, rather than the typical "cap-and-trade". During this period, the government sold allowances to covered enterprises at a fixed price. In Phase 2 (July 2015–July 2018), fixed pricing is replaced by float pricing, and "cap-and-trade" is truly carried out.

During the fixed-pricing period, the carbon emissions regulated by AU ETS accounted for two-thirds of Australian total emissions. About 300 covered enterprises—each with annual CO_2 emissions above 25,000 tons—are mostly distributed in the sectors of stationary energy, industrial process, volatiles, and waste. Natural gas retailers are also included, and they are allowed to transfer their

Table 2.1 Categories of companies that receive free allowances

Levels of emission intensity	Intensity	Percentage of free allocation (%)	Compliance companies
$\geq$ 2000 tCO_{2e}/ mln AUD	High	94.5	Glass, methanol, aluminum smelting
1999–1000 tCO_{2e}/ mln AUD	Medium	66	Lead–zinc, high-purity ethyl alcohol, polyethylene, urea
$\leq$ 1000 tCO_{2e}/ mln AUD	Low	0	Aluminium oxide, coal mine, lime, LNG, gasoline, paper, ethylene

emissions obligation to large-scale natural gas users based on the "Obligation Transfer Numbers" (OTN). The Australian government has been levying liquefied fuel tax upon the users. The large-scale liquefied fuel users may join in the ETS by means of joining in emissions pricing mechanism or paying for liquefied fuel tax.

During the fixed-pricing period, the Australian government divided the covered enterprises into three categories based on their carbon intensity (CO_2 emissions/ monthly business income), and allocated allowances based on the benchmark of varies industries. See company categories and their allowances in Table 2.1.

During the float-pricing period, the companies with high-carbon intensity and cost hard to pass on to end users will continue receiving free allowances, yet the specific allocation method is yet decided. The paid allowances will be allocated through auctions.

The fixed price was initially set as A$23 t/$CO_2$, which will go up 2.5% year on year by taking account of the inflation rate. The Australian government sets a ceiling price that it should not be higher than global prices by A$20 t/$CO_2$. During the float-pricing period, the government sets a floor price at A$15 t/$CO_2$ in 2015–2016 to guard against drastic price fluctuations, such price will rise 4% year on year afterward.

The Australian government has unveiled a slew of supplementary measures to ease the adverse impact of the ETS on socioeconomic growth. First, extending vigorous support to the export-oriented and emissions-intensive industries during the fixed-pricing phase. Second, supporting electric power sector (including the clean energy-based power grid), subsidizing to close outmoded power plants, and allocating free allowances to the large-sized power plants that may be affected. Moreover, in order to mitigate the impact upon electricity consumers, at least 50% of the allowances revenue shall be used for subsiding households, and the amount of subsidy will gradually increase.

In August 2012, Australia and EU announced to link their carbon markets, and achieve complete bidirectional linkage by 2018. At that time, the Australian companies may purchase allowances from EU; the Australian government shall adjust the current carbon pricing policy; calling off the floor price; and updating the offset rules. The coverage of linkage involves the MRV, market supervision, and support to the industrial sectors that are prone to be affected by carbon leakage.

2.1.4 ETS in Japan

Japan took an early start in carbon emissions reduction action. To date, Japan has developed diverse GHG emissions reduction programs. After the KP was concluded, Japan—one of the Annex I Parties—sets an overall reduction target to lower 2012 emissions by 6% from 1990 levels. Japan did not build a cap-and-trade system, but embarked on voluntary reduction.

(1) Keidanren Voluntary Emissions Action Plan (KVAP)

In 1997, Keidanren—the most important and influential business federation in Japan—kicked off the Keidanren Voluntary Emissions Action Plan (KVAP), which regulates massive companies in fields such as manufacturing, energy, transportation, construction, and foreign trade. The scope of regulation extended from the preliminary 38 industrial sectors to 50 industry associations, 1 conglomerate, and 7 railroad companies in 2007. Each industrial sector and company defines the emission reduction and energy saving objectives on their own. The KVAP does not include allowance allocation and trading, it is a voluntary action initiated by companies.

Through a self-assessment, Keidanren acknowledged the accomplishments of the KVAP in lowering carbon emissions and energy intensity and promoting low-carbon energy use. The covered enterprises are on the way of cutting emissions, and some of them have effectively lowered their emissions from 1990 levels. Yet, some nonprofit environmental organizations are skeptical about the reduction effect of the KVAP, because the KP requires the Japanese industrial sectors to cut the 2008–2012 emissions by 8.3% from 1990 levels, while the KVAP target is to cut 2008–2012 emissions to 1990 levels, which is far lower than the KP targets.

(2) Japan Verified Emission Trading Scheme (J-VETS)

In 2005, the Japanese Ministry of the Environment launched the Japan Verified Emissions Trading Scheme (J-VETS) with the jurisdiction over the small-to-medium sized companies that are not covered by the KVAP. The government encouraged companies to send applications, and then picked out the eligible ones based on their reduction cost proposals. The selected companies would receive certain subsidies for installing emissions reduction equipment; meanwhile, they shall bear emissions reduction obligations and fulfill reduction goals through allowance trading. There were altogether 303 companies joining in this program in 2006–2010, their allowances were allocated after their 2002–2004 annual average emissions were calculated and verified.

After the J-VETS was carried out, the annual allowance trading was only at 10,000 tons level on average, i.e., 82,624 t/CO_2 in 2006, 54,643 t/CO_2 in 2007 and 34,227 t/CO_2 in 2008. In contrast, the covered enterprises outperformed their annual reduction targets: they cut 29% of their emissions in 2006 (base year) despite of the target at 21%; they cut 25% in 2007, higher than the target at 19%;

they cut 23% in 2008, far above the target at 8.2%. The operation of J-VETS shows that small companies have great potentials in emissions reduction.

(3) Tokyo Cap-and-Trade Program

In 2010, the Tokyo cap-and-trade program was officially launched, thus becoming the world first cap-and-trade program designed for commercial operation, and the first city-level cap-and-trade program. The program regulates about 1100 industrial and commercial installations, whose emissions accounted for about 40% of Tokyo's total emission. The main objects of regulation are large installations with fuel, heat, and electricity consumption of at least 1500 ton annually. The Tokyo Cap-and-Trade program has extended its emissions reduction target from the KP objectives: to cut the 2020 emissions by 25% from 2000 levels. This overall target is to be fulfilled through two phases. In Phase 1 (2010–2014): to cut annual emissions by 6% from the base year (the average emissions of 3 consecutive years in 2002–2007). In Phase 2 (2015–2019): to cut annual emissions by 17% from the base year. In light of the grandfathering principle, the allowances given to the covered enterprises are calculated based on the emissions in base year and the compliance factors set by the government (the allowances for two phases are calculated at one time), and allocated to the companies at the start of each compliance period for free (Fig. 2.2).

According to the annual summary report released by the Bureau of Environment of Tokyo, the cap-and-trade program has attained significant results: Tokyo cut 2010 emissions by 13% from the base year. Among the total regulated installations, 64% of them overfilled their 2010 reduction targets (the commercial and industrial installations, respectively, cut emissions by more than 8 and 6%); 26% of them cut emissions by more than 17% which was the Phase 2 reduction target; 71% of them were able to fulfill the reduction target on their own, indicating that the remaining 29% had to purchase allowances. Tokyo adjusted up the 2012–2013 reduction targets to 22%. In 2014, only 10% of the installations failed to achieve their intended targets.

(4) Integrated emissions trading market

The KVAP and J-VETS have been developing side by side and complement each other.

The covered enterprises regulated by different emissions reduction systems may carry out additional emissions reduction projects to obtain VER, and the VER would be valid for trading in Japan's integrated emissions trading market that started operating since 2008. In addition, the domestic VER that is developed on basis of the Japan Verified Emissions Reduction (J-VER) could also be traded in this market, so does the VER based on the KP-recognized CDM. In short, the Japanese allowance market has a large capacity and extensive scope.

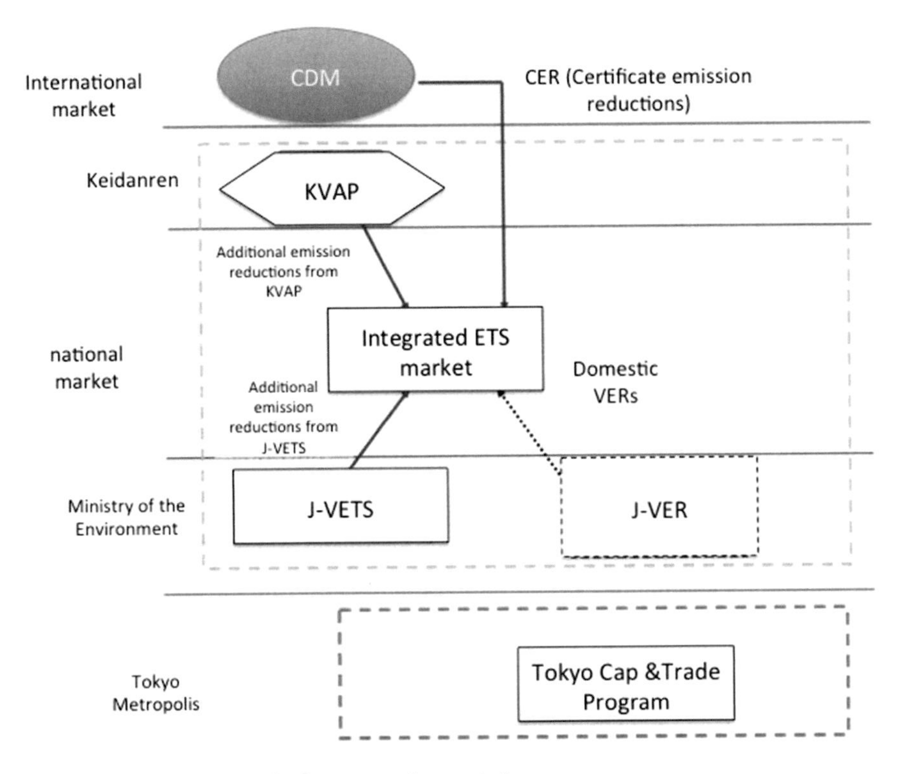

Fig. 2.2 Relations among the Japanese carbon emissions programs

2.1.5 *ETS in Other Countries*

(1) Korea Emissions Trading Scheme (KETS)

The Korean Government came up with the "low-carbon green growth" strategy in 2008, and promulgated the Green Development Law in 2010 which states to fulfill the GHG emissions reduction target in 2020 (emissions decrease 4% from 2005 levels), which has laid a legal basis for carrying out the cap-and-trade system. In January 2015, the Korea Emissions Trading Scheme (KETS) was launched to regulate 68% of the national total emissions, and cut emissions by 37% until 2030, which is equivalent to emission decreasing by 22% based on 2012 level.

There are 525 companies (including 5 domestic aviation companies), which are distributed in the sectors like steel, cement, petrochemical, oil refining, electric power, building, waste, and aviation, under the regulation of the KETS. Six GHGs prescribed in the KP and the indirect emissions from electricity consumption are involved in the KETS.

KETS contains three phases. In Phase 1 (2015–2017), 100% of the allowances are allocated for free, and the quantity of allowances is defined on basis of the companies' average emissions in 2011–2013. The allowances for the cement clinker, oil refining, and aviation sectors are based on benchmarking and the companies' activities in 2011–2013. The government reserves about 5% of the allowances for stabilizing the allowance market and newly operated projects. In Phase 2 (2018–2020), 97% of the allowances will be allocated for free, 3% is left for auctions. In Phase 3 (2021–2025), the free allowances will be no more than 90%, over 10% will be sold in auctions.

The KETS allows for cross-phase allowance banking, yet the allowances are only available for borrowing within the same phase, the amount of lending is no more than 10% of the allowances allocated to the covered enterprises.

During Phases 1 and 2, only the domestic offset credits are allowed for use which are no more than 10% of the total emissions of the covered enterprises. In Phase 2, the projects in other countries may contribute 5% of the offsets. According to the KETS rules, the emissions reduction projects started after April 14, 2010 are compliance projects, and the project owners are noncompliance companies, e.g., the CDM, carbon capturing and storage projects within South Korea. The over-emission companies will be fined three times more than the allowance price.

In order to effectively manage the carbon emissions trading market, the Korean government put into effect the allowance price management mechanism which applies to the following three circumstances [14]:

(i) When allowance prices are three times more than the average prices in past the 2 years for 6 months in a row;
(ii) When allowance prices are two times more than the average prices in the past 2 years for 2 months in a row, and the average trading volume is two times more than the average volume in the same month in the past 2 years;
(iii) When the average allowance prices in any month is 40% lower than the average prices in the past 2 years.

The allowance price management mechanism includes the following measures:

(i) Increase the percentage of reserved allowances up to 25% of the total allowances;
(ii) Develop an allowance regulatory mechanism: no less than 70% and no more than 150% of the total allowances;
(iii) Increase or decrease the ceiling of allowance borrowing;
(iv) Increase or decrease the limit on offset credits;
(v) Set a short-term price ceiling and price floor. The government reserves no more than 25% of the allowances to the new participants.

(2) New Zealand Emissions Trading Scheme (NZ ETS)

In September 2008, New Zealand Parliament adopted the Climate Change Act, which provides for a series of measures to achieve the 10–20% emissions reduction by 2020, and states to establish the New Zealand Emissions Trading Scheme (NZ ETS). In addition to the industrial sector, the NZ ETS also regulates agriculture, fishery, and forestry, but does not set any cap on emissions during the transitory stage. The quantity of allowances is calculated on basis of the companies' industrial output and emissions intensity, and the percentage of free allowances is based on the emissions intensity of companies. The fishery and forestry sectors that are prone to be affected by allowance cost will receive a higher percentage of free allowances. The agricultural sector of New Zealand features high emission and export orientation, so the agricultural companies are also accessible to free allowances. Some of the NZ ETS rules which were effective during the transitory period (from July 2010 to December 2013) still prevail, e.g., the allowances are sold at a fixed price of NZ \$25/t CO_2. The stationary energy, liquefied fossil fuel, and industrial processing sectors are entitled to emit 2 tons of CO_2 with 1 ton of allowance.

During the KP first commitment period (2008–2012), the NZ ETS permit to use offset credits from KP, in order to make the NZ ETS allowance prices synchronize with the global prices.

Some studies show that the emissions reduction cost generated from the NZ ETS may push up electricity and fuel prices, which will, in an indirect manner, raise companies' production cost and household consumption cost. Since the NZ ETS is still implemented in a short time, its impact on cutting GHG emissions is hard to be precisely estimated.

(3) Mexico Emissions Trading Scheme (MX ETS)

In April 2012, the Climate Change Act of Mexico was adopted, which states to cut emissions by 30% until 2020 compared to the Business as Usual (BAU) scenario. Mexican Government was empowered to work on an emissions reduction program, including the building of Mexico Emissions Trading Scheme (MEX ETS). The MEX ETS is made up of two phases: capacity construction and emissions reduction. It was proposed to cover energy production and consumption, transportation, agriculture, forestry, land use, waste disposal, and industrial processing.

(4) India Emissions Trading Scheme (IND ETS)

In 2008, the Indian Government announced the National Action Plan on Climate Change (NAPCC), which states to cut GHG emissions by 20–25% from 2005 levels until 2020. The Indian Government, in addition, to vigorously promote renewables in lieu of traditional energy and raise energy-use efficiency, started operating the energy-related market mechanism, i.e., "Perform, Achieve and Trade" (PAT) and "Renewable Energy Certificate" (REC).

2.2 Gains and Losses of Mainstream Emissions Trading Schemes

The ETS is recognized as an effective mechanism for promoting GHG emissions reduction, lowering emissions reduction cost, and responding to climate change. To date, there are 14 countries either operating or planning for developing the ETS.

The experiences of the states and regions that took the lead in launching the ETS, emissions trading not only motivates emissions reduction, but also generates multiple negative impacts, e.g., carbon leakage and worldwide competitiveness of enterprises. Moreover, the varied design elements of the ETS may generate different effects on the fulfillment of the emissions reduction targets. In order to obtain anticipated effect and weaken the negative effect, it is necessary to analyze the gains and losses of the ETS, conduct assessment of the entire process of the ETS and adjust the mechanism.

The ETS is a market-based environmental policy with one of the aims to promote GHG emissions reduction in countries and regions, with the basic premise of achieving low-cost emissions reduction. Therefore, when assessing the ETS implementation effect, the priority is to discuss whether the compliance entities could deliver abatement and how does the abatement effectiveness; second, whether the abatement cost is relatively lower.

To date, EU ETS is with the longest performance period; the assessments and analyses about this mechanism are the most fruitful. This section will introduce and demonstrate the gains and losses of ETS by quoting relevant literature on the evaluation of operation of EU ETS.

2.2.1 Evaluation of Abatement Effectiveness

ETS is an environmental policy that parallels with other eco-economic policies. The economic development, climate change, production decisions, and non-ETS-driven emissions reduction endeavors may affect the levels of GHG emissions. ETS may overlap with other policies to have abatement effects. For example, energy prices adjustment, subsidizing renewables, and other relevant policies are also have abatement effectiveness. The difficulty faced by researchers is isolating the effect of the EU ETS from other dominant factors. Yet, it is still an arduous task to do so for lack of reliable situational data for comparing with the actual emissions [15].

After reviewing related literature, we find out that the ex-ante analysis method is widely used for calculating the ETS-enabled emissions reduction. Through simulation of scenarios, we can calculate the differences between the emissions reduction under the ETS and BAU.

When making an assessment of the EU ETS emissions abatement effects, Ellerman et al. [16] took account of the historical GDP growth and emissions reduction of the covered enterprises. Through data modeling of 2004 emissions, they made an estimation of the 2005 and 2006 emissions and found out that the emissions reduction of the European 23 countries were about 50 Mt (down 2.4% from 2004); and the cumulative emissions reduction by 2006 was around 100 Mt (down 4.7% from 2004). Moreover, this study compared the prices for carbon, fuel, and gas in Phase 1 and Phase 2 of the EU ETS, and found out that the European electric power generation companies, after shifting from coal-fired power generation to gas-fired power generation, have become a major driver of emissions reduction. In light of the 2004 emissions of the EU members published by the UNFCCC, Ellerman et al. [17] estimated the BAU-based emissions, which show that the EU ETS Phase 1 emissions reduction was about 210 Mt (down 3.5% from 2004). Anderson et al. [18] imported the historical emissions data from the European Statistics Agency into the Dynamic Panel Data Model, adopted energy prices, level of industrial economic activity and climate factor as main conditions, built a model that manifests the relations between various factors and carbon emissions, then estimated the EU ETS BAU-based emissions in 2005–2006. The result shows the EU ETS Phase 1 emissions are around 247 Mt. The comparison between the modeling calculation and the EU ETS verified actual emissions shows that the BAU-based results are overestimated, they shall be around 174 Mt (down 2.8% from 2004). Based on the conclusions made by Ellerman et al. Egenhofer et al. [19] modified the assumptions for the economic activities in order to accord with the actual situations in the EU ETS Phase 2, quantified the emissions in the prior 2 years in Phase 2, and found that contribution ratio of the EU ETS to emissions reduction rose from 1% annually in 2006–2007 to 3.35% annually in 2008–2009.

Based on the econometric model, Murray et al. [20] studied the emissions reduction effect of the RGGI, and found out that the contribution ratio of economic recession to emissions reduction is merely 1%, while the one-third emissions cut of the natural gas market is attributed to the transition from coal-fired power generation to mixed fuels-based power generation. Once the aforesaid variable elements are controlled, the RGGI becomes the major driver of emissions reduction, yet it is still hard for further quantization resolution of the RGGI allowance prices and auction procedures, etc., the accurate research result calls for further studies.

The post evaluation of the ETS effects is based on the micro-data from the companies, yet it is hard to directly gather the corporate data and information, but resorting to the ETS registration system and competent administrations. Wagner et al. [21] by consulting with to the energy and fuel consumption data of each ETS regulated company, calculated the CO_2 emissions from the covered enterprises and found out their emissions were 26% lower than those nonregulated emitters in 2007–2010.

2.2.2 *Evaluation of Reducing the Abatement Cost*

It is a necessity to evaluate the ETS cost-effectiveness, i.e., whether it is able to fulfill the emissions reduction targets at a lower cost. An analytical thinking goes like this, adopting a pre-analysis method, simulating both ETS- and BAU-based scenarios, and comparing their marginal costs for fulfilling the same emissions reduction targets. Whether the ETS is able to realize low-cost emissions reduction is a focal point to be considered and evaluated by the government before the program is launched.

Capros and Mantzos [22] built a PRIMES for the European energy market to compare with the KP-based emissions reduction costs, and observe the changes on the overall emissions reduction cost along with expanding the scope of covered enterprises. The comparison shows that a wider transaction scope could save the expenses on emissions reduction. In case of an ETS-free scenario, the emissions reduction cost holds 0.095% of EU's GDP; such proportion drops to 0.06% in the ETS-based scenario; and it will be as low as 0.025% in a globally industry-wide trading.

The Institute for Prospective Technological Studies (IPTS), one of the seven joint research centers under the European Commission, built a POLES[2] model to compare the emissions reduction costs among all EU countries [23]. The study result shows that the Northern European countries bear more cost for emissions reduction, which is about 0.48% of their GDP; such proportion in other states, like Italy, is generally no more than 0.17%. Although the KP-based emissions reduction targets have increased expenses in this regard, the ETS scenario does bring more benefits to almost all EU countries than the BAU scenario, e.g., the Southern European countries cut their emissions reduction cost by 62%, particularly Germany (50%) and Italy (20%).

The actual carbon market is influenced by supply and demand of emissions allowances (allowances allocation in primary market and allowances trade in the secondary market), competition among traders and asymmetric information. The actual carbon market is not as efficient as the hypothetical market, and the actual carbon prices differ from the theoretically calculated emissions reduction cost, which complicates the post evaluation of the ETS role in cutting emissions. The carbon price is generally affected by fuel price, air temperature, and hydroelectric power generation that impact the marginal cost for emissions reduction. However, Hintermann [24] believes that the carbon price at the initial phase of EU ETS may be affected by the large-scale industrial companies (owner of more free allowances), market expectations of the speculators that opt for market hedging, which causes carbon price to deviate from the marginal cost and results in market bubbles. However, it is fairly difficult to quantify such adverse impacts on the role of the EU ETS in emissions reduction.

According to the Porter Hypothesis, strict environmental regulations can induce efficiency and encourage innovations that help to improve commercial competitiveness. In a long-term view, the ETS is able to motivate the high-emission companies to join in technical innovation and put in place new measures for emissions reduction, so as to mitigate the higher opportunity cost arising from inaction. However, the evaluation from the above perspective is neither an easy job, due to lack of public data about the corporate investment into low-carbon assets and low-carbon public welfare activities. Some researchers are forced to shift to qualitative research to lock up the relevant information. Herve-Mignucci et al. [25] investigated the operational performance of the Germany-based companies under

[2]"POLES" is short for Prospective Outlook on Long-term Energy Systems.

the EU ETS regulation, and found out that most of them realized emissions reduction through funding technical upgrading or improving the production process. Yet the emissions reduction is proved to be a by-product instead of their foremost objectives. Martin et al. made a survey of more than 800 manufacturing companies in six European countries, and learned that after joining in the EU ETS, the large-sized companies called off the investment into high-emission power generation plants, but diverted their funds into the Carbon Capture and Storage (CCS) projects, and turn to supporting more small-scale projects with fewer installments.

Löschel et al., in a qualitative approach, studied the investment of Germany's cross-industry companies that are under the EU ETS regulation in the context of fluctuating allowance prices. The study shows that 77% of these companies have invested or improved the production process that favors emissions reduction; 64% of these companies made investment decisions in 2008–2012. However, most companies (89%) admitted that their original intention was to raise efficiency and reduce energy-induced cost, rather than lowering the cost of compliance. Most of the investigated companies, particularly small-sized ones, have been storing allowances for later. The allowances stored by electricity companies are less than those stored by industrial companies, indicating that industrial sectors—one of the external factors—may drive up the demand for emission allowances.

2.2.3 Other Impacts Evaluation

(1) Value redistribution and "windfall profits"

The economic cost arising from the ETS occupies a small share in the GDP, and free from exerting a significant negative impact on the overall economic performance, shown by massive studies. However, the ETS rules and allowances circulation have resulted in the redistribution of emission rights among different countries, industries, and income earners, and divided value distribution. Any incompliance of the allowance allocation principles may lead to uneven distribution of emission rights among countries and "windfall profits" of certain industries, and widen wealth gap. As a result, the efficiency of the ETS has become a widely debated issue across the world.

Take the EU ETS for instance. Owing to the disparity in economic development, different EU countries have varied demand for emission allowances, thus causing frequent allowance circulation and massive transfer of allowance value among these countries. In its Phase 1 implementation, the EU ETS traded about €505 million of allowances [26]. The volume of transferred allowances (from the surplus countries to the deficit countries) reached 650 Mt, which was 11% of the EU's total allowances and worth of €5200 million. The total import-and-export allowances were about 218 Mt, worth of €9.41 billion, mainly flowing from Poland, France, the Czech Republic, and the Netherlands to such importers as the UK, Spain, Italy, and

Germany. The allowance sufficiency also varied among industries. Trotignon and Delbosc [26] revealed that the EU electricity sector was short of 6.1% of allowances, while other sectors were in surplus, e.g., the surplus proportion of cement and paper sectors was 4 and 20%, respectively.

The EU ETS was abiding by free allowance allocation during the prior two phases of implementation, it means that when some industries are short of allowances, they may transfer the allowance cost to obtain windfall profit. The occurrence of windfall profit is a negative outcome of the value distribution of ETS. Several studies have shown that the European electricity sector has passed the allowance cost on to consumers by raising electricity price.

Some studies have shown that the ETS may widen the gap between the rich and the poor. Terry Dinan and Diane Lim Rogers argue that both the auction-based ETS and carbon tax will drive up carbon prices and generate income distribution effect. Feng et al. [27] compared the wealth gap, respectively, induced by the carbon tax and ETS, and concluded that the cost from carbon tax held 6% of the gains of the lowest income earners and 2.4% of the gains of the high-income earners. In case of the ETS-induced cost, it was 4.3% of the gains of the lowest income earners and 1.7% of the gains of the high-income earners.

The transnational allowance circulation and allocation, which directly reveals the fact the industrial sectors in different countries have varied demand for allowances, enables efficient distribution of allowances. However, free allowance allocation fails to fully reflect the real supply–demand pattern of allowances, and may affect the efficiency of allowance allocation. Moreover, different income earners have paid great attention to the increasing cost of carbon emissions, which creates an important basis for adjusting the ETS.

(2) Carbon leakage effect

Carbon leakage occurs when there is an increase in CO_2 emissions outside the country which is under a strict climate policy. In light of the *IPCC Fourth Assessment Report: Climate Change 2007*, integrated carbon leakage rate of one region or country equals to the proportion of CO_2 emissions increase outside the region or country to CO_2 abatement inside the region or country. The two parties shall have trading contacts, resource/energy import or export, or geographic proximity. Leakage can occur through three channels, including: (i) Carbon leakage results from transnational/trans-regional trade. Since the local compliance industries have all along been bearing high-emissions cost, disadvantageous market occupancy, and profit margin, the regulation areas may reduce production and export of emissions-intensive products in the short run, but an increase in the import of these products from nonregulation countries, which consequently transfers the production of such products to the export countries and increases their emissions, instead of altering the global total emissions, carbon leakage is thereby caused. (ii) An increase in local fossil fuel prices resulting, for example, from mitigation policies may lead to the reallocation of production to regions with less stringent mitigation rules (or with no rules at all), leading to higher emissions in those regions and, therefore, to carbon leakage. (iii) Carbon leakage may also arise from technical spillover—cross-regional technical development and promotion, yet such portion of carbon leakage is hard to quantify for unavailable direct data and information.

With Gemini-E3 CGE Model, Bernard and Vielle [28] calculated the integrate carbon leakage rate within the EU ETS framework at about 7%, of which 2.3% is through trans-regional trading. With GTAP-E Model (comparative-static CGE Model) and considering the impacts from the Border Adjustment Measures, Kuik and Hofkes [29] estimated the integrated carbon leakage rate of within the EU ETS framework at about 8.2–10.2%. The carbon leakage rate of the steel sector drops the most vigorously owing to these measures, i.e., dropping from 35 to 29% based on national unified policy; and falls even further to 2% based on the national differentiated policy. FitzGerald et al. [30] ranked the pricing power of different industries in a top-down order, and took account of the impacts from energy price and carbon tax, and finally concluded that the European steel sector is the most vulnerable to have carbon leakage.

(3) Impact on industrial competitiveness

The covered enterprises of each country's ETS are mainly distributed in energy- and carbon-intensive industries. Generally, the companies with low unit production cost, high-profit margin, and a large market occupancy will be more competitive in markets. Under the framework of the ETS, the companies with unit emissions higher than a defined amount of emissions allowances have to pay the additional environmental cost, which increases the production cost of these companies and alters their competitive strength. Moreover, owing to different percentages of free allowances and allowance allocation approaches, the regulation companies will have varied competitive edge in emissions pace and environmental cost.

Based on a scenario of 100% of allowances allocation, if 15 British industries would not cut carbon emissions in the short term, Hourcade et al. [31] then calculated the percentage of emissions cost in their industrial added value (IAV). Through comparison, he found out that both cement and steel sectors have the highest percentage of emissions cost. The German Federal Ministry for the Environment, Nature Conservation, Building and Nuclear Safety (BMU) organized similar studies of the domestic industries, and concluded that the cement, lime, chemical fertilizer, and steel sectors bear the highest emissions cost.

Based on a scenario of free allowances allocation, Quirion and Demaily [32] developed the CEMSIM Model[3] and GEO Model[4] for analyzing the impact of grandfathering allocation and benchmarking allocation on the European cement output and profit. Based on grandfathering allocation, even if the allowances are 50% of the historical emissions, the cement sector is still profitable as usual, the

[3]The CEMSIM Model is developed to assess the energy and technical development trends of the European cement industry.

[4]The GEO Model is developed to assess the effect of carbon emissions trading.

cement output declines remarkably accompanied with serious carbon leakage. In contrast, based on benchmarking allocation, the allowances are no more than 75% of the historical emissions, the cement sector is moderately affected with both output and profit dropping more than 5%.

Through analyzing the differentiated impact of the EU ETS on the marginal production cost of the British different industries, Oxera finds out that the industrial marginal production cost will increase along with rising emission allowances prices, there is also value appreciation of the free allowances granted to the industry at the same time, which may make up for the loss incurring from rising marginal production cost. Thus, the selection of allowance allocation approach is fairly important.

2.2.4 Conclusion and Adjustment to ETS

The ETS is of great significance in helping a society achieving the carbon emissions reduction targets, and effectively lowering the emissions reduction cost. A review of the EU ETS operation experiences shows that the prior two phases of operation generated 50–100 Mt of emissions reduction annually. In order to avert from any excessive allowances allocation as in Phase 1, more allowances were auctioned in Phase 2. In contrast to the BAU-based emissions reduction, the ETS motivates covered enterprises to fulfill emissions reduction target at a lower cost. Meanwhile, the investigations of the EU companies reveal that the ETS triggers the high-emissions companies to update emissions reduction technologies, adopt measures and work out investment plans that favor emissions reduction, in an aim to lower the potential high opportunity cost arising from inaction in the future.

The impact of the ETS on a society varies among different regions and industries. In reference to the EU ETS experiences, most of Phase 1 and 2 allowances were subject to free allocation; the electricity sector passed on the allowance costs and other costs onto the consumers, which widened the wealth gap and garnered windfall profit. The integrated carbon leakage rate arising from the EU ETS is about 7–10%. The steel and aluminum sectors are less competitive and the most vulnerable to the foreign rival companies; the cement sector is fairly competitive and free from serious impact from the ETS.

In order to tackle the above problems arising from the EU ETS operation, the EU plans to adjust and reform the scheme during Phase 3, in an aim to revitalize the weak emissions trading and prevent from excessive allowance allocation within the EU.

In addition to helping an area achieve low-cost emissions reduction, the ETS is, in fact, using allowances prices as a signal to call for companies to divert investment into low-carbon technologies, and guard against the "lock-in effects" [33] from continuous funding high-carbon technologies—the great expectation of the

Table 2.2 Strength and weakness of three allowances supply management mechanisms

Type of mechanism	Description	Strength	Weakness
Economic activity-based	Regulate allowances supply in light of the changes in the macroeconomic indicators, like GDP, primary energy consumption, petroleum price, and relevant financial market indicators that reflect commodity market development	The economic indicators are usually public and available data, and they are objective enough to prevent SMM from affected by subjective policies and decisions. The economic indicators also exhibit the direct correlation between commodity market and allowances market, thus making participants make more accurate judgment about carbon market	The macroeconomic indicators only represent the overall social economic activities, rather than the actual economic performance of the ETS covered enterprises. The data incompleteness, lagged updating, or inconsistency may affect the implementation effect of SMM
Allowances surplus-based	Regulate supply of the allowances surplus (verified yet not auctioned). Set a lower threshold for the allowances surplus, if it is above the threshold, then withdraw the allocated allowances from the carbon trading market	Such regulation, which is transparent, simplified, and comprehensible, enables the ETS to flexibly cope with the emergent economic shocks and inrush of international VERs	The transaction behavior of the ETS-based market players is guided by the future allowances policy; therefore, it is difficult to define the instant lower threshold for the allowances surplus, which, in turn, affects implementation effect of SMM
Allowances price-based	Regulate allowances supply in light of carbon price. Set lower threshold for carbon price, if it is above the threshold, then withdraw the allocated allowances from the carbon trading market	Develop an explicit and definite price signal, based on which the companies will limit their compliance cost within a designated price range, and explore low-carbon technologies at the lowest cost, which is favorable for companies to making investment decisions. Open and transparent data about carbon prices, basis for SMM, provide market players with more clear-cut bases for decision-making	The price-based allowances supply mechanism is deemed as distorting the connotation of carbon prices reflecting emissions reduction cost, and affecting the price discovery principle, which may lead to carbon price fluctuations between price ceiling and price floor, and make SMM like a dynamic carbon tax system. The varied price ranges will hold up the linkage between different trading mechanism

EU placed on the ETS. Therefore, in case of any extreme situations taking place on the carbon market, it is necessary to extend interventions to maintain carbon prices at a reasonable level which is favorable for the fulfillment of the emission reduction targets.

Kollenberg and Taschini [34] holds that the current EU ETS policy measures are incapable of market feedback, so the changing EU economic environment will lead to extreme uneven allowances allocation, yet the ETS itself fails to draw any effective market feedback. Although the European governments decided not to allocate the verified 900 Mt allowances before 2020, it could only resolve the superficial problem as low carbon prices in the short term, rather than essentially enabling the ETS to be more capable of market feedback. Such one-off measure is neither able to handle the increasingly complicated economic fluctuations in the future, nor alter the corporate expectations for carbon price decline. Besides, the repeated raising or lowering of the percentage of the allowances allocation is unworkable within the EU policy framework. The amount of allowances should not be decided by the governments on their own, but through gaming and consultation among multiple stakeholders.

The building of an ETS feedback mechanism is, therefore, a necessity. Its essential role is to alter the corporate intrinsic pessimistic expectation of carbon prices, and turn to adjusting allowances supply flexibly in light of economic environment, and guide market players to keep a close watch on carbon price trend, make corresponding strategies and finally develop a virtuous cycle. A feedback mechanism is able to encourage regulation companies to store up allowances in case of low carbon prices to boost up the prices; and sell out allowances in case of high carbon prices to adjust down the prices.

An allowances supply management mechanism shall have the following core functions:

(1) Increase elasticity of the ETS to deal with external shocks or extreme events, and improve the policies to be more adaptable to time dimension and desired effect;
(2) Improve the accuracy of the allowances supply adjustment plan, e.g., explicitly prescribe when and how the adjustment shall be made;
(3) Exempt from any policy discretionary power in addition to market rules so as to avert from possible policy intervention.

Currently, there are three allowances supply management mechanisms: economic activity-based mechanism, allowances surplus-based mechanism, and allowances price-based mechanism. Each mechanism has its own strength and weakness, and capable of managing allowances supply in light of real supply and demand. Their comparison is shown in Table 2.2.

References

1. Wenjun Wang. UK's Climate Change Policy and Its Lessons [J]. Contemporary International Relations, 2009 (09): 29-35.
2. Commission Stuff Working Paper-Impact Assessment[R]. SWD (2012) 177 final, Brussels: European Commission, 2012.
3. Minsi Zhang, Di Fan, Yong Dou. Analysis of EU ETS Operation Progress and Enlightenment on China [EB/OL]. (2014-02-12). http://www.ncsc.org.cn/article/yxcg/yjgd/201404/20140400000848.shtml.
4. European Commission Decision of 29.3.2011 [S]. C (2011)1983 final.
5. Commission Regulation (EU) NO. 176/2014 of 25 February 2014 [J]. Official Journal of the European Union, 2014 (NO. 176/2014).
6. COMMISSION REGULATION (EU) No 5502011 of 7 June 2011 [J]. Official Journal of the European Union, 2011 (No. 550/2011).
7. Commission Staff Working Document Exclusive Summary of the Impact Assessment [R]. SWD (2015) 136 final, Brussels: European Commission, 2015.
8. Customs Tariff Amendment (Carbon Tax Repeal) Bill 2014 [S].
9. RGGI 2012 Program Review: Summary of Recommendations to Accompany Model Rule Amendments [R]. Final Program Review Materials, U.S: RGGI Inc, 2013.
10. Second Control Period Interim Adjustment for Banked Allowances Announcement [R]. General Documents, RGGI Inc, 2014.
11. Regional Greenhouse Gas Initiative Memorandum of Understanding [J]. 2005.
12. Final Regulation Order [S].
13. Yan Wen, Changsong Liu, Yong Luo. Review and Analysis on U. S. Carbon Emissions Trading System [J]. Advances in Climate Change Research, 2013, 9 (2): 144–149.
14. Korea Emissions Trading Scheme [R]. International Carbon Action Partnership, 2016.
15. LAING T, SATO M, GRUBB M, et. al. The effects and side-effects of the EU emissions trading scheme: The effects and side-effects of the EUETS [J]. Wiley Interdisciplinary Reviews: Climate Change, 2014, 5(4): 509–519.
16. ELLERMAN A D, BUCHNER B K. Over-Allocation or Abatement? A Preliminary Analysis of the EU ETS Based on the 2005–06 Emissions Data [J]. Environmental and Resource Economics, 2008, 41 (2): 267–287.
17. ELLERMAN A D, CONVERY F J, DE PERTHUIS C. Pricing Carbon [M]. The United States of America by Cambridge University Press, New York, 2010.
18. ANDERSON B, DI MARIA C. Abatement and Allocation in the Pilot Phase of the EU ETS [J]. Environmental and Resource Economics, 2011, 48 (1): 83–103.
19. EGENHOFER C, ALESSI M, GEORGIEV A, et al. The EU Emissions Trading System and Climate Policy Towards 2050-Real Incentives to Reduce Emissions and Drive Innovation [R]. Brussels: Centre for European Policy Studies (CEPS) Brussels, 2011.
20. BRIAN C. MURRAY, PETER T, MANILOFF, EVAN M. MURRAY. Why have GHG in RGGI States Declined-An Econometric Attribution to Economic, Energy Market, and Policy Factors [R]. Working Paper EE 14-01, U.S: Nicholas Institute for Environment Policy Solutions, Duke University & Colorado School of Mines & Trinity College of Arts and Sciences, Duke University.
21. MARTIN R, MUÛLS M, WAGNER U J. The Impact of the EU ETS on Regulated Firms: What is the Evidence After Eight Years? [J]. Available at Social Science Research Network 2344376, 2014 (31).
22. European Union energy outlook to 2020 [M]. CAPROS P, ETHNIKON METSOBION POLYTECHNEION. Ms completed on 30 September 1999. Luxembourg: Off. for Off. Publ. of the Europ. Communities, 1999.
23. Preliminary Analysis of the Implementation of an EU-wide Permit Trading Scheme on CO2 Emissions Abatement Costs [R]. Institute for Prospective Technological Studies (IPTS), 2000.

24. RALT MARTIN, MIRABELLE MUÛLS, ULRICH WAGNER. An Evidence Review of the EU Emissions Trading System, Focusing on Effectiveness of the System in Driving Industrial Abatement [R]. United Kingdom: Department of Energy & Climate Change of UK, 2012.
25. ANDREAS LÖSCHEL, BODO STURM, REINHARD UEHLEKE. Revealed Preferences for Climate Protection when the Purely Individual Perspective is Relaxed-Evidence from a Framed Field Experiment [R]. Discussion Paper No. 13-006, Centre for European Economic Research (ZEW), Mannheim, 2013.
26. RAPHAEL TROTIGNON, ANAIS DELBOSC. Allowance Trading Patterns During the EU ETS Trial Period: What Does the CITL Reveal? [R]. Climate Report No 13, Mission Climate of Caisse des Dépôts, 2008.
27. FENG K, HUBACEK K, GUAN D, et al. Distributional Effects of Climate Change Taxation: The Case of the UK [J]. Environmental Science & Technology, 2010, 44 (10): 3670–3676.
28. BERNARD A, VIELLE M. Assessment of European Union Transition Scenarios with a Special Focus on the Issue of Carbon Leakage[J]. Energy Economics, 2009, 31: S274–S284.
29. KUIK O, HOFKES M. Border Adjustment for European Emissions Trading: Competitiveness and Carbon Leakage [J]. Energy Policy, 2010, 38(4): 1741–1748.
30. FITZGERALD J, KEENEY M, SCOTT S. Assessing Vulnerability of Selected Sectors Under Environmental Tax Reform: the Issue of Pricing Power [J]. Journal of Environmental Planning and Management, 2009, 52 (3): 413–433.
31. HOURCADE J-C, DEMAILLY D, NEUHOFF K, et al. Climate Strategies Report: Differentiation and Dynamics of EU ETS Industrial Competitiveness Impacts [R]. Climate Strategies, 2007.
32. DEMAILLY D, QUIRION P. CO2 Abatement, Competitiveness and Leakage in the European Cement Industry Under the EU ETS: Grandfathering Versus Output-based Allocation [J]. Climate Policy, 2006, 6 (1): 93–113.
33. RALF MARTIN, MIRABELLE MUÛLS, ULRICH WAGNER. Climate Change-Investment and Carbon Markets and Prices-Evidence from Manager Interviews [R]. Climate Policy Initiative, 2011.
34. SASCHA KOLLENBERG, LUCA TASCHINI. The European Union Emissions Trading System and the Market Stability Reserve-Optimal Dynamic Supply Adjustment [R]. University of Duisburg-Essen & Grantham Research Institute of London School of Economics, 2015.

Chapter 3
Overview of Chinese Pilots ETS and Characteristics

For the purpose of fulfilling China's carbon emissions reduction targets by 2020 through a cost-efficient market mechanism, and expediting transformation of economic growth pattern and upgrading industrial structure, China's State Council, at the end of 2011, issued the *Work Plan for Greenhouse Gas Emissions Control during the 12th Five-Year Plan Period*, which requires to "explore and establish a national unified carbon emissions trading market." In response to the State Council's plan, the NDRC, in October 2011, initiated a carbon emissions trading pilot program in seven regions, including five municipalities as Beijing, Tianjin, Shanghai, Chongqing and Shenzhen, and two provinces as Hubei and Guangdong, the earliest pilot ETS which open its carbon market is Shenzhen ETS in June 2013, and the latest launched is Chongqing ETS in June 2014. According to NDRC's statistics, more than 1900 emitters (companies and institutions) are covered in this pilot ETS, receiving total emission allowances at around 1200 $MmtCO_{2e}$. During the pilot period, the seven pilots ETS have completed the whole procedures, i.e., data collection, a stipulation of rules, compliance mechanism, offset rules, etc. Each pilot ETS designed its proposal according to the local situation, which is the basis of Chinese national ETS. The structure and characteristics of these pilots ETS are described as follows:

3.1 Construction and Operation of China's Pilots ETS

China's seven pilots ETS which are distributed in the eastern, central and western regions, involve three administrative levels (province, municipality, and city). The basic framework of China's pilots ETS is similar to that of other countries, but distinguishes between regions in details, such as coverage scale, allowance cap, coverage standard, etc. (see Table 3.1).

In reference to the *Interim Measures for the Administration of Carbon Emission Permit Trading* and relevant policy schemes, the pilots ETS are distinguished in

© China Environment Publishing Group Co., Ltd. and Springer Nature Singapore Pte Ltd. 2019
D. Zhao et al., *A Brief Overview of China's ETS Pilots*,
https://doi.org/10.1007/978-981-13-1888-7_3

Table 3.1 Comparison among Chinese pilots ETS

Location	Start date	Criteria for coverage scope	Coverage percentage in total emissions	Allocation plan	Compliance rule	Trading administration
Beijing	Nov 28, 2013	The fixed installations (e.g., iron and steel, cement, petrochemical) that directly or indirectly emitted an annual average of 10,000 $mtCO_{2e}$ or more in 2009–2011	–	The annual total emission allowances are allocated to both established and newly operated facilities and also for adjustment; 5% are reserved for competent department	The compliance rate reached 97.1% in 2013 and 100% in 2014. The percentage of carbon offsets is no more than 5% of the year's total allowances, and that of local CCER is no less than 50%	Restrict overly stocking up allowances; set up price early-warning system; the government maintains the stability of carbon market through allowances auction or buyback
Shanghai	Nov 26, 2013	The industrial sectors that emitted 20,000 $mtCO_{2e}$ or more (e.g., iron and steel, petrochemical, chemical, nonferrous, electricity) or nonindustrial sectors that emitted 10,000 $mtCO_{2e}$ or more in any year in 2009–2011	In 2013: about 57% of Shanghai total emissions	The allowances are allocated for free. Three years' allowances are allocated to covered enterprises in a one-time manner with the benchmarking approach	The compliance rate was the same as 100% in 2013 and 2014. CCER, though accepted for compliance, should be no more than 5% of the year's total allowances	Shanghai Environment and Energy Exchange, which trades three categories of allowance products (SHEA13, SHEA14, and SHEA15), mainly relies on listed transaction and sometimes exercises negotiating transfer, with competitive bidding playing a supplementary role. A price rise or drop by no more than 30% is tolerated. A risk management system is established

(continued)

Table 3.1 (continued)

Location	Start date	Criteria for coverage scope	Coverage percentage in total emissions	Allocation plan	Compliance rule	Trading administration
Tianjin	Dec 26, 2013	The major emitting sectors (e.g., iron and steel, chemical, electricity, heat, petrochemical, oil and gas exploitation) or civil construction field that emitted an annual average of 20,000 $mtCO_{2e}$ or more since 2009	Holding 50–60% of Tianjin total emissions	The allowances are allocated for free. The allowances granted to electricity and heat sectors are based on benchmarking allocation and subject to post-correction, while the allowances to industrial sectors are based on grandfathering allocation and adjusted in light of pre-phase reduction effect and targets. Different allocation approaches apply to established and newly operated facilities	The compliance rate reached 96.5% in 2013 and 99.1% in 2014. CCER, though accepted for compliance, should be no more than 10% of the year's total allowances	The allowances transaction is realized through online spot trading, trade by agreement, or auction trading. A price rise or drop by no more than 10% is tolerated. A risk warning system (setting a cap on the quantity of allowances) and an auditing system are established
Chongqing	June 19, 2014	Metallurgy, electricity, chemical, building materials, and machinery sectors, and light industrial sectors	About 60% of Chongqing total emissions	The allowances are allocated for free. Three years' allowances are allocated to covered enterprises in a one-time manner. The companies shall compete for allowances and the government shall set a cap on total allowances; in other words, the government may review and adjust the allowances applied by the companies	The compliance rate in 2014 reached 70% (as of July 14, 2015). The covered enterprises shall not use carbon offsets unless there is short of allowances, and the percentage of offsets may not be above 8% of the allocated allowances during the compliance period	Emission allowances and CCER are two tradable products. The covered enterprises are imposed by certain restrictions, i.e., the allowances to be sold may not be above 50% of their annual total allowances. A price rise or drop by no more than 20% is tolerated

(continued)

Table 3.1 (continued)

Location	Start date	Criteria for coverage scope	Coverage percentage in total emissions	Allocation plan	Compliance rule	Trading administration
Guangdong	2013.12.19	The industrial sectors that emitted 20,000 $mtCO_{2e}$ or more (or consumed energy in 10,000 tons of coal equivalent) in any year in 2010–2012, e.g., electricity, cement, petrochemical and iron and steel	In 2013: about 54% of Guangdong total emissions	The 2014 allowance allocation altered greatly from the 2013 version, i.e., only a certain proportion of allowances were allocated to the covered enterprises for free based on their verified quantity of allowances; the established and newly operated covered enterprises shall meet different allocation criteria; both the grandfathering and benchmarking allocations are applicable	The compliance rate reached 98.9% in 2013 and 100% in 2014. CCER, though accepted for compliance, should be no more than 10% of the year's total allowances	The covered enterprises, before their actual annual emissions are verified by GD DRC, may not transfer over 50% of the year's free allowances in their registered account into their transaction account for trading. Neither an investment institution nor individual may hold allowances more than 3 Mmt CO_{2e}. A price rise or drop by no more than 20% is tolerated. And a price stabilizing and reserving mechanism is established
Hubei	2014.04.02	The 12 industrial sectors that consumed energy in no less than 60,000 tons of coal equivalent in any year in 2010–2012, e.g., iron and steel, chemical, and cement	In 2013: about 35% of Hubei total emissions	The allowances are calculated with the grandfathering approach and allocated for free (the allowances granted to electricity sector are based on the benchmarking principle). A review and accreditation mechanism for allowance allocation is established	During the first compliance period (ending in July 2015), Hubei-based covered enterprises fulfilled 100% compliance. CCER, though accepted for compliance, should be no more than 10% of the year's total allowances	An emission allowances trust system is built. The government reserves 8% of total allowances for guarding against market risks. No more than 30% of the allowances reserved by government are used for Price Discovery

(continued)

Table 3.1 (continued)

Location	Start date	Criteria for coverage scope	Coverage percentage in total emissions	Allocation plan	Compliance rule	Trading administration
Shenzhen	2013.06.18	Top 800 companies in rank list of industrial added value, top 4,000 electricity-consuming companies, major oil-burning companies and boiler-operating companies. These covered enterprises are defined through cross comparison guided by four principles	In 2013–2015: about 38% of Shenzhen total emissions	The allowances are allocated for free. The covered enterprises will, through competition, receive 3 years' allowances in a one-time manner	The compliance rate reached 99.2% in 2013 and 99.7% in 2014. CCER, though accepted for compliance, should not be above 10% of the companies' actual emissions in the previous compliance period	Carry out the systems for full-amount trading, restricting price rise/drop and maximum allowances holding, big emitters' reporting, mandatory reduction of allowances holding, supervising block trade, monitoring abnormal conditions, risk warning, risk disposal fund, and important customer information disclosure

Source 1. *Interim Measures for the Administration of Carbon Emission Permit Trading*, *Carbon Emissions Trading Implementation Plan* and other relevant policies and documents issued by local governments, or posted on the websites of local carbon emissions exchanges
2. Pang et al. [1]
3. Wang et al. [2]

five aspects[1]: coverage scope, allowances cap, allocation plan, transaction administration, and compliance mechanism.

With respect to coverage scope, China (Shenzhen) Emission Exchange (CEEX) has the widest coverage scope. In addition to the industrial and construction sectors, CEEX has taken the public transportation into its ETS. In terms of total allowances, Guangdong ranks first among seven pilots in 2013 and 2014, respectively, reaching 388 and 408 Million ton CO_2, accounting for around 50% of the aggregated allowances of all pilots ETS. Moreover, Guangdong exerted the policy of auction to distribute the allowances first. In terms of carbon market vitality, China Hubei Emission Exchange is a pioneer in carbon financing innovation and allowances transaction. As of October 30, 2015, Hubei Emission Exchange had traded 23.71 $MmtCO_{2e}$, receiving turnover of 570 mln yuan; its carbon market trading scale has ranked the first among seven pilots ETS in both trading volume and turnover; moreover, new breakthroughs were made in carbon asset pledging and inviting overseas investment.[2] While speaking of the administration of carbon emissions trading, Shanghai Environment and Energy Exchange (SEEX) and Shenzhen Emission Exchange (CEEX) are prominent examples. SEEX took the lead in coming up with an emissions accounting guideline and accounting methodology for the covered enterprises. Shenzhen—a special economic zone vested with legislative power—was the first in enacting a local decree for carbon emissions trading, which was known as the *Provisions of Carbon Emissions Administration of the Shenzhen Special Economic Zone*.

Overall, the above analysis provides us a visualized comparison between the seven carbon markets: they have the same architectural mechanism, but vary in specific rules, which may result from their different locations and policy orientations. For instance, in a region in which its economy heavily relies on the secondary industry, its ETS coverage focuses on industrial companies. A region with the sound market economy basis and mature financial environment, its ETS works hard on more elaborate administration rules. For the ETS that cares more about carbon emission reduction cost, carbon financing innovation is emphasized. In a word, each pilots ETS has its own characteristics.

---

[1]Currently, China's carbon trading schemes remain in the pilot period, the concerned areas have been slow in the legislation work. Despite of Shenzhen, the other six areas have been working on their *Temporary Measures for Guangdong Carbon Emissions Management*. All areas have stipulated elaborate technical specifications on Monitoring, Reporting, and Verification (MRV) of the covered enterprises. MRV is an independent factor for building the ETS, and the ETS in different areas have diversified coverage scope, so it is of limited meaning in comparing local MRV in either horizontal or vertical manner.

[2]China Hubei Emission Exchange. Hubei Carbon Market Watch (Oct. 2017) [EB/OL]. http://www.hbets.cn/jbXwzx/2332.htm.

3.2 The Pilot ETS in Five Municipalities

China authorized seven provinces/municipalities to carry out the ETS pilot program. Owing to different administrative levels, the municipal and provincial ETS take account of different factors in system design. They are introduced successively in Sects. 3.2 and 3.3.

3.2.1 Beijing Pilot ETS

Beijing Pilot Carbon Emissions Trading Scheme (briefed as "Beijing pilot ETS") covered CO_2 emissions from all fixed installations within the administrative jurisdiction of this municipality.

(1) Content and procedures of administration

During the pilot period, Beijing pilot ETS constraints covered the following emitters within the administrative jurisdiction of Beijing Municipality: direct CO_2 emissions from the fossil fuel combustion of fixed installations, industrial production process, waste disposal, etc., as well as indirect CO_2 emissions from electricity consumption of fixed installations [3]. Two categories of companies fall into the regulation of Beijing pilot ETS: covered enterprises and reporting companies. The covered enterprises refer to the key emitters with annual total CO_2 emissions (direct and indirect) at or above 10,000 tons; they are obliged to control their CO_2 emissions during the compliance period, which is called as the covered enterprises. The reporting companies, with annual comprehensive energy consumption at or above 2000 tons of standard coal equivalent but below 10,000 tons, voluntarily receive ETS regulation (in reference to the regulation upon key emitters) and report their emissions during the compliance period to the ETS administrator.

The covered enterprises shall, before each April 5 during the pilot period, submit a hard-copy emissions report and verification report to the ETS municipal administrator, which will carry out review and spot check of the two reports. The administrator shall also introduce measures for the administration of the verifier, lay down requirements for the verifier's recording conditions and monitoring duties, and exercise dynamic administration of the verifier through an on-site inspection or irregular spot check.

On September 1, 2014, Beijing Municipal Government released the *Beijing Carbon Emissions Offset Administration Measures* (*for trial*), which authorizes the covered enterprises use CCERs, emission reductions from energy-saving projects and from forestry carbon sequestration projects to offset part of their CO_2 emissions. According to the measures, 1 ton of verified emission reductions in carbon dioxide equivalent is able to offset 1 ton of CO_2 emissions; and the offsets may not be above 5% of the year's allocated allowances. There are elaborate provisions on the proportion of the offsets from CCERs, emission reductions from energy-saving projects and from forestry carbon sequestration projects. For instance, the CCERs from any offset project outside Beijing may not be more than 2.5% of the year's allocated allowances; the CCERs from Beijing-based offset projects shall be over 50%; and the western region-based offset projects are the first choice when looking for nonlocal offsets.

The reporting companies shall deliver the reports on their last year's emissions to the Beijing ETS administrator in the first quarter (no later than March 20) of each year since 2014. The covered enterprises shall fulfill their compliance in the second quarter (no later than end of June) of each year since 2014: specifically, the covered enterprises shall deliver a verification report, apply for additional allowances for last year's newly added facilities and for allowances adjustment before April 5; their application will be approved before April 30; the current year's allowances to the established facilities shall be allocated before June 30; the covered enterprises shall fulfill their compliance (paying off last year's allowances) before June 15.

(2) Reward and penalty mechanism

In accordance with Beijing municipal laws, any covered enterprises that fail to perform their obligations as reporting, monitoring, or verification will be punished. Beijing imposes harsher punishments upon the violators, i.e., a fine that is 3–5 times more than the market average carbon price will be imposed on the overdue emission part of allowances. In order to ensure administrative enforcement of law, Beijing specially provided for the *Provisions on Standardizing Administrative Penalty Discretion over Carbon Emissions Trading* (No. 1 [2014] BJ DRC), which was the first one in doing so among the seven pilot carbon markets.

In order to guard against any potential market manipulation, Beijing pilot ETS is equipped with two firewalls as control of emission allowances inventory and trading price.

(i) Restrict overly stocking up

For the covered enterprises, their maximum allowances inventory may not be over the sum of its annual allowances and 1 million tCO_{2e}. For the reporting companies, their maximum allowances inventory may not be over 1 million tCO_{2e}. For any natural person that intends to join in emission allowances trading, their maximum allowances inventory may not be over 50,000 tCO_{2e}.

(ii) Price early warning

In order to regulate market activity, Beijing Development and Reform Commission (BJ DRC) conducts open market operation through allowances auction mechanism or buyback. No more than 5% of the year's total allowances are reserved for auction. If the daily weighted average price for the allowances is above 150 yuan/tCO_{2e} for 10 consecutive days, BJ DRC will organize interim allowances auction to keep down carbon prices. In case such price is below 20 yuan/tCO_{2e} for 10 days in a row, BJ DRC shall, through negotiations with the municipal finance department and financial supervision administration, decide whether to buyback the allowances, and the buyback quantity, price, and approach, and then send the repurchase instruction to Beijing Climate Change Research Center (BCRC).

(3) Cross-regional cooperation mechanism

In mid-December 2014, Beijing Municipal Development and Reform Commission, in collaboration with Hebei Provincial Development and Reform Commission and Chengde[3] Municipal Government, released the *Notice on the Matters about Promoting Cross-regional Pilot ETS Program* (No. 2645 [2014] BJ DRC), marking an official start of the Beijing–Hebei joint carbon market. As a result, both the offsets projects in Beijing and Hebei are recognized by Beijing ETS.

The Beijing–Hebei joint carbon market is the first cross-regional emission reduction endeavor in China. The covered enterprises in Chengde City of Hebei province will be treated equally by Beijing ETS. Other qualified institutions and natural persons in Chengde are allowed to trade in Beijing ETS. Chengde-based offset projects as valid as those in Beijing. Moreover, Beijing–Hebei joint carbon market gives priority to developing forestry carbon sink. On September 24, 2014, Shunyi Forestry Carbon Sequestration Project (Phase 1) was listed on the website for trading in CBEEX. It is part of the plain forestation project launched by Beijing Shunyi District Landscape and Forestry Bureau. The project, which covers an afforested area of 9452.2 mu (630 ha)[4] within Shunyi District, pre-issued offsets of 1995 tCO_2. On December 30, Chengde City Fengning County Qiansongba Forestry Carbon Sequestration Project (phase 1)—the first offset project that crosses Beijing and Hebei—was listed for trading in CBEEX; it sold offsets of 3450 tCO_2 and received turnover of 131,000 yuan on the same day. As of January 16, 2015, this project had accumulatively sold offsets of 15,000 tCO_2, and earned a total turnover of more than 570,000 yuan (averaging at 38 yuan/t), proving that Beijing and Hebei achieved success in cross-regional eco-compensation through a market-oriented approach.

3.2.2 *Tianjin ETS*

In 2013, Tianjin Municipal Government issued the *Implementation Plan for Carrying out the Work for Tianjin Pilot Carbon Emissions Trading Scheme* and the *Temporary Measures for Tianjin Carbon Emissions Administration*, which lay out specific instructions for carrying out the ETS pilot work in Tianjin [4]. These two documents explicitly announce that Tianjin will establish the Cap-and-Trade (CAT) system and the CAT-based emissions trading scheme, key CO_2 emitting sources reporting system and emissions verification system. On December 24,

[3]Chengde is one of the cities in Hebei Province and a northeastern city in China. Chengde borders on Beijing at the southwest.

[4]1 ha = 15 mu.

2013, Tianjin Development and Reform Commission (TJ DRC) released the *Notice on Carrying out the Work of Carbon Emission Trading Pilot Program*, which requires the relevant institutions to do a good job in emissions monitoring, report delivery, emissions, and allowances administration, together with such appendixes as the industrial emissions accounting and reporting instruction, allowances allocation plan and registration system operations guide.

(1) Content and procedures of administration

In light of China's *Standard Industrial Classification* (SIC) and the statistics about the key energy consumers since 2009, Tianjin ETS regulates five major emitting industrial sectors as iron and steel, chemical, electricity and heat, petrochemical, oil and gas recovery, as well as the civilian construction companies with annual emissions above 20,000 tCO_2. Based on the emissions verification results, Tianjin selected 114 companies as the covered enterprises among the above five sectors, their combined annual emissions account for 50–60% of Tianjin total, which also proves that Tianjin GHGs emissions are relatively concentrated. During the pilot period, the GHG regulated by Tianjin ETS will be carbon dioxide.

In light of the *Temporary Measures for Tianjin Carbon Emissions Administration*, Tianjin ETS will mainly rely on free allowances allocation, which is supplemented by allowances transaction (auction or sales at a fixed price). The funds that are generated from allowances trading will be spent for special purposes, like the work for controlling GHGs emissions.

(i) Allowances administration

In December 2013, Tianjin made public the *Allowances Allocation Plan for the covered enterprises under Tianjin Carbon Emissions Trading Scheme (for trial)*, which proposes to allocate allowances by following the Grandfathering and Benchmarking principles, and take account of the companies' annual emissions reduction targets, competitiveness, energy-use efficiency, reduction efforts before joining the ETS, and industrial baseline emissions.

According to the provisions of Tianjin pilot ETS, the covered enterprises may transfer their annual allowances into the following year of compliance until May 31, 2016. If the covered enterprises dissolve, close down, or move out of Tianjin, they shall clear the allowances that are equal to their actual CO_2 emissions during their operation in the year of compliance, and turn over the year's remaining free allowances. In case any company is incorporated into the covered enterprises, the former shall inherit the latter's allowances and corresponding rights and duties. In case a regulated company is divided, it shall draw up a plan for dividing its allowances and obligations, and submit the plan to Tianjin Development and Reform Commission, and then re-register for amended allowances.

In case any change occurs in the covered enterprises' organizational structure owing to incorporation, division, or dissolution, they shall submit the relevant materials and documents to the competent department to explain such change, the department shall study the reasons for such change and then decide to either transfer or recover the allowances granted to these companies.

In terms of the validity period of the allowances, the covered enterprises may carry forward their non-canceled allowances to the following year of compliance until May 31, 2016. After that, the validity period of the allowances is subject to the relevant state regulations.

Tianjin Climate Exchange (TCE) trades emission allowances by means of online trade, contractual trade, and auction-based trade. In online trade, the object of the transaction is the emission allowances coded as TJEA; the transaction volume is in a unit of 10 tons or an integral multiple of 10 tons, and the minimum offer price is in the unit price of RMB (yuan)/tCO_2, with the minimum price movement at 0.01 yuan/tCO_2. The trading system acts in accordance with the principle of "price-time priority". Besides, in order to control trading risks, TCE has established the full-amount settlement system, price limit system, large trader reporting system, position limit system, risk early-warning system, risk reserve system, and auditing system. The price fluctuation rate may not be above 10%. TCE has specially set a risk control division to improve the internal administration, safeguard the interests of investors, and ensure safe and steady performance the of carbon market.

(2) Carbon offsetting

The covered enterprises under Tianjin ETS are allowed to use a prescribed proportion of CCERs to offset their emissions, i.e., the offsets may not be more than 10% of their actual emissions in the year. One CCER is to offset 1 ton of CO_2 emission, and the CCERs may come from any project, any area, or region. On March 24, 2015, Tianjin Tianfeng Steel Co., Ltd. and China Carbon Futures (Beijing) Asset Management Co., Ltd. made the first deal to purchase 60,000 tons of CCERs through TCE online trading.

(3) Characteristics of Tianjin ETS

As compared with other pilot schemes, Tianjin ETS registration takes on two characteristics: (i) it recommends three options for allowances clarify, i.e., compliance clear, voluntary clear, and mandatory clear. The compliance clarify is an action for the covered enterprises to fulfill their emissions reduction commitment. The voluntary cancelation is an option limited to nongovernment users. In terms of mandatory clarification, the competent department may cancel the allowances account of the covered enterprises if it fails to fulfill the reduction commitment, until its allowance account is empty. Both compliance and voluntary clarify are carried out by the covered enterprises on their own, while the mandatory clarify is operated by the ETS system administrator authorized by the government.

(ii) The allowances shall be transferred from register account to transaction account before the transaction. Such design is to mitigate risks from misoperation.

3.2.3 Shanghai ETS

After the NDRC announced to carry out the pilot ETS pilot program, Shanghai Municipal Government spent some time on preparation, and then issued its ETS implementation plan and opinions [5], which explicitly clarify the guiding principles, crucial factors, job schedule, and so on.

(1) Content and procedures of administration

When choosing the industries that covered by the ETS regulation, each pilot area takes account of the local economic development level and the demand for industry transition in the future. Take Shanghai for instance, since it is now at the transitional stage of post industrialization, it not only regulates the CO_2 emissions from industrial sectors (iron and steel, petrochemical, chemical, nonferrous, electricity, building materials, textile, paper, rubber and chemical fiber, etc.), but from non-industrial sectors (aviation, port, airport, commerce, hotel, finance, etc.).

With respect to the ETS participants, Shanghai Environment and Energy Exchange (SEEX) only allowed the covered enterprises to join in the trade at the opening stage. Since September 2014, after SEEX issued the *Measures for the Carbon Trading Market to Implement the Institutional Investor Eligibility System (for trial)*, and the *Account Opening Guide for Institutional Investors*, Shanghai ETS was officially open to common institutional investors.

As for allowances calculation, most pilot areas comply with the Grandfathering Principle (in reference to local industrial characteristics and historical emissions), which is supplemented by the Benchmarking Principle. Shanghai ETS adopts the Benchmarking Principle widely in such sectors as electricity, aviation, port, and airport. Hubei chooses the Grandfathering Principle, but switches to the Benchmarking Principle for computing additional allowances to electricity companies.

As for allowances allocation, Shanghai allocates most of the allowances for free, but the government usually reserves a certain proportion of allowances for auction in order to help companies' compliance or tackle with abnormal market phenomena.

Table 3.2 Shanghai ETS reward and penalty rules

Allowances adjustment	Review the validation mechanism and the provisions for the covered enterprises' closure or removal
Risk control	Establish the price limit system (the price fluctuation rate be no more than 30%), position limit system and risk reserve system
Default penalty	Urge the covered enterprises to pay off their allowances, and concurrently impose a fine of no less than 50,000 yuan but no more than 100,000 yuan

(2) Carbon offsetting

The covered enterprises under Shanghai ETS may use the CCERs to offset 5% of their total allowances or trade such as CCERs on the trading administration platform. The trading administration platform is made up of the registration system, transaction platform, and MRV platform. All trading activities are held in Shanghai Environment and Energy Exchange (SEEX). As for the policy documents, SEEX has unveiled not only elaborate transaction rules, but detailed requirements for member administration, transaction settlement, information management, etc.

(3) Reward and penalty mechanism

SEEX has set several supplementary mechanisms like allowances adjustment, risk control, and default penalty, in an aim to handle any special or abnormal market phenomena (Table 3.2).

3.2.4 Chongqing ETS

The emitting sources in Chongqing are unevenly distributed, i.e., the emissions from the central urban area are relatively higher, while the emissions from the county area are much lower. Such factor shall be taken into account for building the ETS in Chongqing.

(1) Content and procedures of administration

Chongqing ETS is the only one of the reducing GHG emissions in total but not the intensity reduction differed from the rest six pilots ETS, with the benchmarking allocation as a supplement. Both emissions offsetting and banking are acceptable.

When choosing the covered enterprises for Chongqing ETS, several key conditions are taken into account, i.e., energy-use efficiency, large-sized companies with emissions reduction potentials or large-sized companies in rectification industries. Chongqing ETS regulates the industrial companies with annual CO_2 emissions above 20,000 tCO_{2e} in any year in 2008–2012 (based on the energy consumption of more than 10,000 tce), including such sectors as electricity, metallurgy, chemical, and building materials. Six types of GHGs (CO_2, CH_4, N_2O, HFCs, PFCs, and SF_6) are subject to the regulation [6].

The competent department allocates free allowances to the covered enterprises, but the amount of allowances for the existing emitter differs from that for newly emitter, time boundary point is on December 31, 2010. In the case of carbon emissions from the existing emitter, the baseline allowances are the maximum annual emissions over 2008–2010, then decrease year by year since 2011. In case of the newly emitter, the baseline allowances are the maximum annual emissions of the past 3 years before joining the ETS, then decrease year by year. All the annual allowances before 2015 are allocated all at once.

Chongqing ETS has two types of the allowance allocation, the allowances granted to the existing emitter are different from the newly emitter. For the covered enterprises with existing emitter, Chongqing ETS sets a strict cap based on their history emissions; for the covered enterprises with newly emitter, the allocation proposal is in line with their baseline emissions.

Chongqing ETS market trades allowances, CCERs, and other eligible products authorized by the Chinese central government, which are quantified in the unit of tCO_{2e} and priced in the unit of yuan/tCO_{2e}. The difference between Chongqing ETS and other pilot areas is that the former limits the amount of annual tradable allowances, which may not be more than 50% of the annual free allocated allowances.

(2) Carbon offsetting

Chongqing ETS recognizes such flexible mechanisms as offsetting and banking. The covered enterprises may use the emissions reductions from local offset projects (e.g., carbon sequestration), which are outside of the coverage scope for offsetting part of their emissions, and the offsets may not be more than 8% of the total allowances for each compliance period. The offset projects shall start operation after December 31, 2010 (excluding carbon sequestration projects) and fall into one of the following categories: energy saving and higher energy-use efficiency, using clean energy and non-hydro renewable energy, carbon sequestration, energy-based activities, industrial production process, agriculture, and waste disposal. The allowances are available for banking, i.e., the covered enterprises may transfer the surplus allowances in the current compliance period to the next period.

In contrast to other pilot areas, Chongqing ETS has its own requirements for the offsetting mechanism: (i) the acceptable CCER offsets are no more than 8%, which lies at the middle level among the seven pilot areas. (ii) Nonlocal offset projects are also recognized. (iii) Chongqing set restrictions on the operation time, project type, and location of the projects that contribute CCERs. The above requirements reflect that Chongqing ETS competent department pays great attention to CCER, which could be interpreted as a stance to motivate more and more local companies to join in clean energy industries.

Chongqing is the only pilot area that permits offsets can be used just in the condition of existing allowances gap, which implies that if the total allowances and allowances allocation remain unchanged, there is no allowances gap, the usage of CCER will not be allowed. It should be noted that Chongqing does not recognize hydropower projects as CCER sources. The above regulations will enable Chongqing carbon market to become a de facto single allowances market.

(3) Reward and penalty mechanism

In order to incentivize the covered enterprises to fulfill their emissions reduction targets, Chongqing gives priority to supporting these companies to improve emissions administration; support them undertake such projects subsidized by the Central Government budget, e.g., the comprehensive demonstration of fiscal policies to promote energy conservation and emissions reduction, as well as resource saving and environmental protection; the municipal special funds for energy conservation and emissions reduction shall be first diverted to the covered enterprises; encourage financial agencies to provide green financing services to the covered enterprises.

In terms of penalty, the *Decisions on the Matters about Carbon Emissions Administration of Chongqing Municipality (Exposure Draft)* state that if any covered enterprises failed to deliver emissions report or rejects verification according to the ETS regulation rules, the competent department shall order it to make corrections within a definite time; otherwise, it shall be fined by 20,000–50,000 yuan. In case any regulated company refuse to pay off the allowances or underperforms such obligation, the competent department shall, in reference to the emissions that are beyond the allowances, impose a fine that is three times more than the average allowances trading price in the month ahead of the compliance deadline. If a third-party verification agency is found to issue a false verification report, the competent department shall impose a fine of 30,000–50,000 yuan to the unqualified agency.

3.2.5 Shenzhen ETS

The economy of Shenzhen relies on the tertiary industry, the major CO_2 emitters refer to the industrial, construction, and transportation sectors, Shenzhen ETS covered 635 industrial companies and 194 public buildings, they can be divided into three categories in light of their industrial property: (i) there are 13 public institutions which produce and supply of water, natural gas, and electricity. (ii) There are 15 large-scale enterprise groups with high output value, e.g., Huawei Technologies, Zhongxing Telecommunication Equipment (ZTE), and Foxconn. (iii) Other manufacturing companies. The covered enterprises in the construction sector are made up of the government office administrations, public buildings with sole proprietorship, and property management companies [7 and 8].

(1) Content and procedures of administration

In light of the *Implementation Plan for Carrying out the Work for Shenzhen Pilot Carbon Emissions Trade Scheme* issued by Shenzhen Government, the industries/institutions that are covered by Shenzhen ETS shall satisfy the following criteria: (i) The companies/public institutions with annual emissions above 5000 tCO_{2e}. (ii) The large public buildings with a floor area above 20,000 m^2 or the government office buildings with a floor area above 10,000 m^2. (iii) Other companies, public institutions, or buildings that apply to join in the ETS voluntarily and receive approval by the competent department. The covered emission of Shenzhen ETS accounted for 38% of Shenzhen's total emissions in 2010.

(i) Allowances setting

Shenzhen ETS adopts the combination approaches of the top-down and bottom-up method to set the total amount of allowances in reference to Shenzhen's emissions reduction targets, GDP growth predictions, and emissions reduction potentials of industrial sectors.

Shenzhen sets a cap based the consideration both of total amount of emission control and on emissions intensity. To set the allowances for the covered enterprises in light of the expecting emission intensity reduction targets and anticipated output, it is favorable for adjusting the allowances to deal with any drastic economic fluctuation. However, such an approach complicates the operation of the carbon trading system, and increases management cost upon the competent department and the covered enterprises.

(ii) Allowances allocation [9]

Shenzhen ETS allowances are based on free allocation and paid allocation, the latter is achieved through auction or fixed-price sales.

With respect to the allowances allocation system, Shenzhen ETS prefers pre-allocation that computes and adjusts the allowances based on the IAV of the covered enterprises. In order to guarantee the fairness and rationality of allowances allocation, Shenzhen ETS pre-allocates the allowances to the industrial sectors that abide by the Benchmarking Allocation, i.e., in light of the allocation criteria, allocate the free allowances to the covered enterprises in advance, when the companies' actual emissions have been verified, the competent department shall, in reference to the verification results

With respect to the allocation methodology, Shenzhen ETS allocates the allowances based on the Benchmarking Principle. For a sector with a single product, the pre-allocated allowances are the companies' carbon intensity target multiplied by their anticipated output. For a sector with nonsingle product, the pre-allocated allowances are the companies' carbon intensity target multiplied by IAV.

In order to define the carbon intensity target of an individual company, Shenzhen proposes an allowances allocation mechanism based on the theory of Bounded Rationality in Repeated Games, i.e., categorize the companies with similar industry type, product type, and business scale into the same group, the government sets a cap on the total allowances granted to different groups, then ask the companies of the same group to report their 2013–2015 emissions and IAV targets concurrently via a game software; the allowances will be auto computed and pre-allocated to them based on default rules. If the companies refuse to accept the outcome, they may repeat the above process again and again until they cease altering the emissions and IAV data for reporting. At that time, multiply the emissions by IAV (ultimately accepted by the companies), the outcome will be the companies' carbon intensity target. Such methodology increases proactivity of the companies in allowances allocation, improves administration efficiency, introduces human intervention in allowances allocation which may guard against such malpractice as power rent-seeking. However, such methodology is quite complicated, the categorization of the company calls for massive basic data, the loss or inaccuracy of the basic data (e.g., IAV) will hold back application of such methodology.

(iii) Allowances administration

Under the framework of Shenzhen ETS, each compliance period for the covered enterprises is 1 calendar year; the allowances surplus in the previous year may be banked for use in the coming year. Shenzhen ETS grants the allowances for 3 years at one time, so it especially provides that the allowances issued for the coming year may not be used for the previous year's compliance (Table 3.3).

(2) Carbon offsetting

Shenzhen ETS allows the covered enterprises to use the CCERs to offset no more than 10% of their emissions. See, Shenzhen ETS provisions on the offset proportion and project type in Table 3.4.

Table 3.3 Characteristics of Shenzhen ETS allowances administration

Allowances banking	The allowances left from the previous year are banked for use in the coming year
Allowances loaning	The allowances issued for the coming year may not be used for previous year's compliance
Allowances shortage	Make up the shortage before May 30 each year
Allowances surplus	When the covered enterprises move out, announce bankruptcy or dissolution, if the pre-allocated allowances are more than 50% of their fulfilled compliance obligation, such allowances shall be recovered by the competent department; otherwise, the surplus allowances are disposed of by the covered enterprises
Allowances buyback	The allowances bought back by the competent department every year may not be more than 10% of the year's valid allowances
Allowances use	The covered enterprises are entitled to assign or pledge their allowances in accordance with law
Allowances adjustment	Based on the covered enterprises' actual output (the sector with the single product) or IAV (manufacturing sector) in the previous year, confirm their actual allowances, then supplement or deduct the pre-allocated allowances
Allowances settlement	Each calendar year

Table 3.4 Shenzhen ETS carbon offsetting mechanism

General guideline	If the CCERs are generated by the covered enterprises/institutions within their permissible emissions scope, such CCERs may not offset their emissions
Acceptable CCER proportion	The CCER offsets may not be higher than 10% of the covered company's actual emissions in the previous year
Valid CCER offset projects	The valid offset projects for Shenzhen carbon market fall into five categories: renewable energy and new energy (wind power, solar energy generation, waste incineration power generation, rural household biogas, and biomass power generation), clean transportation, marine ecosystem carbon sequestration, forestry sequestration, agricultural emissions reduction

Table 3.5 Shenzhen ETS reward and penalty mechanism

Reward for compliance	The covered enterprises shall strictly abide by the administration rules; any company that achieves greater reduction results will be commended or rewarded by Shenzhen Municipal Government
Fines for noncompliance	The concerned company will be imposed a fine that is three times more than the average market price for the portion of extra emissions; with financial subsidy abolished and in ccessible to any municipal financial assistance within 5 years
Bad credit rating	Access to the social credit management system and make public the noncompliance companies
Implementation	Four noncompliance companies were strictly sanctioned in 2013

(3) Reward and penalty mechanism

Shenzhen ETS has worked out strict and elaborate rules to punish any violation or noncompliance of the covered enterprises and institutions by a heavy fine and a bad credit rating. In contrast to other ETS pilot areas, Shenzhen stresses financial penalties which are fairly deterrent, if the concerned companies will be fined for extra emissions, their financial assistance will be ceased. For instance, four covered enterprises failed to fulfill their obligations on time during the 2013 compliance period, the competent department, in light of the *Interim Measures for the Administration of Shenzhen Carbon Emissions Trading*, report their noncompliance to the corporate social credit administration and the financial credit management agency, made public the names of these companies in line with relevant regulations, and reported their noncompliance to the financial department which ceased their financial assistance (Table 3.5).

3.3 Introduction to the Pilot ETS in Two Provinces

Hubei and Guangdong are the only provinces of ETS among the seven ETS pilot areas. Hubei lies in central China, while Guangdong is located in the south. They are different from each other in economic development level, industrial structure,

CO_2 emissions level, and social low-carbon awareness. Meanwhile, they share such common features as uneven local economic growth, arduous emissions reduction task, complete industrial structure, and great disparity in emissions structure. In a word, the ETS pilot work in Hubei and Guangdong is fairly representative.

3.3.1 Hubei ETS

Hubei is a major province in central China, covering an area of 185,900 km^2. It contributed a GDP of 2.95 trillion yuan in 2015, the permanent population reached 58.52 million at 2015 end, and the year's energy consumption totaled 164.04 Mtce [10]. Hubei is now at the fast-growth track. After the NDRC announced to involve Hubei in the ETS pilot program, the province specially made some preparations, and then the Provincial Government issued the *Implementation Plan for Carrying out the Work for Hubei Pilot Carbon Emissions Trade Scheme*, which explicitly provides for the main principles, essential elements, and work progress for carbon trading. Before the carbon market was officially open to business, a government directive was issued to introduce the interim measures for carbon trading administration, which will support the implementation of the ETS pilot program.

(1) Content and procedures of administration

Regulated industrial sectors: Hubei is now at the phase of rapid industrialization, the major emitters under the ETS regulation are mostly the industrial sectors.

Allowances calculation: Most pilot areas adopt the Grandfathering Principle that relies on the characteristics and historical emissions of industrial sectors, the Benchmark calculation is only a supplement. Hubei ETS also gives priority to the Grandfathering Principle, while the new electricity plants abide by the Benchmark principles. Most of the allowances are allocated for free, yet the competent department may reserve a certain proportion of allowances for auction (to assist companies in their compliance) or deal with any abnormal market situations. In light of the *Hubei ETS Allowances Allocation Plan*, the allowances reserved by the government shall be no more than 8% of the total allowances; no more than 30% of the allowances reserved by the government are used for carbon price discovery; any non-tradable allowances surplus or reserved allowances surplus shall be revoked.

Carbon market participants: The covered enterprises, and the legal person institutions, other organizations and individuals that join in the ETS voluntarily (Table 3.6).

(2) Carbon market performance

By the end of March 2015, China Hubei Emission Exchange (CHEX) had been in operation for almost 1 year, finishing more than 230 dealing days. Its total trading volume occupied about 43% of the aggregate trading volume of the seven

Table 3.6 Characteristics of Hubei ETS design

Start date	April 2, 2014
Criteria for covered enterprises	The industrial companies with comprehensive energy consumption at or above 60,000 tce in either 2010 or 2011
Regulated industrial sectors	Twelve industrial sectors, e.g., electricity, iron & steel, cement, chemical
Allowances calculation	Both Grandfathering and Benchmarking Principles are applicable to electricity sector. The Grandfathering Principle is followed by all the other sectors
Allocation approach	Free allocation. The initial allowances for a compliance period are allocated to the covered enterprises (after registration) at one time
Allowances administration	The total allowances are made up of the allowances to established companies, the allowances to newly operated companies and those reserved by government The allowances reserved by government shall be no more than 8% of the total allowances. No more than 30% of the allowances reserved by government are used for price discovery
Deadline of compliance period	The last workday of each May
Main body	The covered enterprises, and the legal person institutions, other organizations and individuals that join in the ETS voluntarily
CCER	10%

pilot carbon markets, and the total turnover was about 30% of the gross turnover of the seven markets.

In contrast to Shanghai Environment and Energy Exchange, CHEX encountered robust transactions shortly after it was open to business, then the trading activities gradually cooled down and become stable, but showed drastic fluctuations as approaching the end of 2014. In reference to such characteristic developments, the operation of CHEX could be split into three phases: (i) Phase 1: from April to July 2014. (ii) Phase 2: from August to November 2014. (iii) Phase 3: from December 2014 to March 2015. The trading share in these three phases accounted for 59.4, 14.9 and 25.6% of CHEX total trading volume, respectively.

The strike price in CHEX has been lower than the other six carbon markets. From the opening in April 2014 to the end of March 2015, CHEX reported an average striking price at 24 yuan/tce, which was fluctuating moderately from 22 to 26 yuan/tce.

(3) Reward and penalty mechanism

China Hubei Emission Exchange has set several supplementary mechanisms like allowances adjustment, risk control, and default penalty, in an aim to handle any special or abnormal market phenomena (Table 3.7).

Table 3.7 Hubei ETS reward and penalty rules

Allowances adjustment	Review the validation mechanism and the provisions for covered enterprises' closure or removal
Risk control	The government reserves 8% of the total allowances
Default penalty	Impose a fine that is 1–3 times (no more than 150,000 yuan) on the current year's average carbon price upon the extra emissions. Double quota deduction for the part of over emission in the next year's allowances

3.3.2 Guangdong ETS

Guangdong is located at the southernmost tip of China's mainland, covering an area of 179,700 km^2. Guangdong is China's most developed province, it contributed a GDP of 7.28 trillion yuan in 2015, the permanent population reached 108 million at 2015 end, and the year's final energy consumption totaled 256.62 Mtce. Guangdong is now at the mid-to-late phase of industrialization, the tertiary industry accounted for 50% of the provincial GDP; it is an example for the other developed regions in China.

In terms of Guangdong ETS, its system design and operational experiences are worthy of learning and popularizing, its institutional organization and administrative accountability have offered inspirations and references for carbon emissions reduction. Therefore, an all-round interpretation of Guangdong ETS and the pilot program in other areas is of far-reaching impact on pressing ahead with the construction of a nationwide unified carbon market.

(1) Content and procedures of administration

The CO_2 emissions in Guangdong mainly come from industry and manufacturing, therefore, industrial sectors become the foremost regulated objects. The electricity, cement, iron, and steel and petrochemical sectors came under Guangdong ETS regulation in the first compliance period.

In light of the *Implementation Plan for Carrying out the Work for Guangdong Pilot Carbon Emissions Trade Scheme* issued in 2012 [11], all the industrial companies, which are located within the administrative areas of Guangdong, emitted CO_2 at or above 10,000 t (or consumed comprehensive energy at 5000 tce) in any year over 2011–2014 are defined as "reporting companies", i.e., they do not take allowances administration or join in allowances trading, but report their emissions to the government every year. The industrial companies, which are located within the administrative areas of Guangdong, emitted CO_2 at or above 20,000 t (or consumed comprehensive energy at 10,000 tce) in any year over 2011–2014 are "covered enterprises", which are distributed in electricity, cement, iron and steel, ceramics, petrochemical, textile, plastics, paper sectors, etc. In 2013, the electricity, cement, iron and steel, and petrochemical sectors were first came into the

ETS regulation, their CO_2 emissions in the year accounted for around 60% in Guangdong total emissions.

With respect to the allowances allocation system, Guangdong ETS integrates both free allocation and paid allocation; the latter is achieved through auction or fixed-price sales. Guangdong ETS prefers pre-allocation that computes and adjusts the allowances based on the actual production of the covered enterprises. In order to guarantee the fairness and rationality of allowances allocation, Guangdong ETS pre-allocates the allowances to the industrial sectors that abide by the Benchmarking Allocation, i.e., in light of the allocation criteria, allocate the free allowances to the account of the covered enterprises, when the companies' actual emissions have been verified, the competent department shall withdraw or give more allowance according to the allocation standard.

With respect to the allocation methodology, Guangdong ETS allocates the allowances based on the Benchmarking Principle; make the average emissions intensity of industries as the baseline, then raise the baseline year by year through the annual reduction factor, so as to tighten up total allowances and drive companies to cut emissions. Since Guangdong ETS total amount of allowances and allocation plan are open and transparent, such methodology may provide an explicit and stable anticipation to the companies.

Under the framework of Guangdong ETS, each compliance period for the covered enterprises is 1 calendar year; the allowances surplus in the previous year may be banked for use in the coming year (Table 3.8).

Table 3.8 Characteristics of Guangdong ETS allowances administration [12–14]

Allowances banking	The allowances left from the previous year are banked for use in the coming year
Allowances loaning	N/A
Allowances shortage	Make up the shortage before June 30 each year
Allowances surplus	If the covered enterprises intend to shut down or move out, they shall, before completing the relevant formalities, settle the allowances based on their actual and verified emissions in the year; the free allocations for the non-production months (the operation rate in the month is below 50%) will be recovered by the GD DRC. The allowances surplus are disposed by the covered enterprises
Allowances purchase	The newly comer enterprises shall purchase full-amount allowances according to the estimated emission at the bidding platform
Allowances use	Before the GD DRC verifies their annual emissions, the covered enterprises may not transfer over 50% of the year's free allowances in their registered account into their transaction account for trading
Allowances adjustment	The actual allowances based on the covered enterprises' actual output in the previous year, then supplement or deduct the pre-allocated allowances. The companies whose production is above their capacity by no more than 30% are recognized according to the relevant regulations
Allowances settlement	Each calendar year

(2) Carbon offsetting

Guangdong ETS allows the covered enterprises to use CCERS (no more than 10% of the year's total allowances) to offset their emissions. See the characteristics of Guangdong carbon offsetting mechanism in Table 3.9.

(3) Reward and penalty mechanism

Guangdong ETS has worked out elaborate provisions on reward and penalty of the covered enterprises and verification institutions. Guangdong ETS usually executes the penalty by imposing fines and lowering credit rating (Table 3.10).

Table 3.9 Guangdong ETS carbon offsetting mechanism

General guideline	If the CCERs are generated by the covered enterprises/institutions within their permissible emissions scope, such CCERs may not offset their emissions
Acceptable CCER proportion	The CCER offsets may not be higher than 10% of the compliance company's actual emissions in the previous year; no less than 70 of 10%, of the CCERs shall come from the local VER projects
Valid CCER offset projects	Non-hydro electricity generation, non-fossil fuel electricity generation and heat supply (not based on coal, petroleum, or natural gas, but CBM is unconstrained), waste energy (heat, pressure, and gas) utilization Exclude the CDM projects that have generated reductions before registration at the UN CDM Executive Board
Others	The offsets from the CO_2 and CH_4 emissions reduction projects are valid, and CO_2 and CH_4 reductions shall account for more than 50% of all GHG emissions from these projects

Table 3.10 Guangdong ETS reward and penalty mechanism

Reward for compliance	The covered enterprises that, which have fulfilled their obligations, shall take the precedence to declare the government-subsided projects (incl. low-carbon emissions, energy conservation and emissions reduction, renewable energy, circular economy); shall take the precedence to government financial support
Fines for noncompliance	The unsettled allowances for the next year shall be fined by 2 times, a fine of 50,000 yuan is concurrently imposed
Penalty on credit	The noncompliance is reported to the credit management system of financial institutions and Guangdong social credit system; keep the public informed
Penalty on project	No qualify to apply new projects during the period of ETS trial

3.4 Summary

A unified China carbon market is still at the infant phase. The seven ETS pilot schemes share common ideas, but vary in details (e.g., general design, coverage scope, allowances amount, allowances allocation, offsetting, compliance mechanism, and MRV), owing to diversified political conditions, economic growth, stage of development, and industrial structure. Their similarities and existing problems are concluded as follows:

(1) Select the covered enterprises or institutions mainly based on their energy consumption;
(2) Compute total amount of allowances by combining the "top-down" and "bottom-up" methodologies;
(3) Allocate the allowances to electricity sector based on the "Benchmarking Principle";
(4) All of them have built an effective MRV system and carbon trading platform;
(5) All of them set a cap on proportion of offsets (10%), but the sources of the offsets vary among the ETS pilot areas.

Some problems exist in the current ETS pilot mechanisms:

(1) There are great uncertainties in setting the allowances cap. In terms of the "top-down" approach, it takes account of the GDP growth of the pilot area, it is an uncertain factor and not always the same with the GDP growth of the ETS regulated industries, thus injecting uncertainties into the allowances cap based on the "top-down" computation. In terms of the "bottom-up" approach, it relies on the historical emissions and reduction potentials of the covered enterprises. However, such approach is neither 100% reliable if the pilot area has a fast-developing economy and robust GDP growth, local companies are expanding their capacity year by year, and multiple new companies come into being to generate a rigid increase in emissions.
(2) An extensive coverage scope may lower the economic efficiency of the ETS pilot scheme. Different pilot areas have their own coverage scope, either big or small, some only involve large-sized companies with high emissions and high energy consumption, e.g., iron and steel, electricity, heat, metallurgy, cement, and petrochemical sectors. In contrast, other may regulate as many as 20–30 industrial sectors, including service, large public buildings, and public institutions (colleges and government agencies). During the current pilot period, the ETS coverage scope should not be too extensive, otherwise, it will inevitably increase the difficulties and costs in allowances allocation, MRV execution, and companies' compliance.
(3) The allowances allocation based on the Grandfathering Principle does not fit the ETS mechanism in the rapidly developing areas. The grandfathering allocation is based on the historical emissions of the covered enterprises, which may conflict with the demand for sustainable economic growth; and the industry prosperity cycle may alter the emissions structure, which will result in

a gap between allowances supply and demand. Take Beijing for instance, in around the base year, the iron and steel and cement sectors were sluggish, many companies halted production; but the electricity generation plants were operating at a full load, implying that more allowances shall be granted to the electricity sector. However, after the economy entered into the New Normal phase, the allowances supply–demand pattern in these sectors reversed, thus altering the production cost of the covered companies.

(4) The way of linkage between the current pilots ETS and the future national ETS shall be considered. Pilots ETS vary from each other in allowance setting, MRV rules, allocation approaches, it is hard to homogenize under the different value of allowances.

(5) The carbon market functions are fairly weak. As compared with other mature capital markets, e.g., the securities market, China's pilot carbon markets are not strongly functional, particularly weak in mobility, since it is still young and trading newborn product.

References

1. Tao Pang, Li Zhou, Maosheng Duan, Study on the Linking of China's Emissions Trading Pilot Schemes [J]. China Population, Resources and Environment, 2014, 24 (9): 6–12.
2. WANG,Wenjun, XIE Pengcheng, LI Chongmei, LUO Zhigang, ZHAO Daiqing. The key elements analysis from the mitigation effectiveness assessment of Chinese pilots carbon emission trading system [J]. China population, resources and environment, 2018, 28(4): 26–34.
3. People's Government of Beijing Municipality. Pilot measures for Beijing's carbon emission trading mechanism [EB/OL]. http://www.bjpc.gov.cn/.
4. People's Government of Tianjin Municipality. Pilot measures for Tianjin's carbon emission trading mechanism [EB/OL]. Tijian People's Government website, http://www.tjzfxxgk.gov.cn/tjep/ConInfoParticular.jsp?id=44984.
5. People's Government of Shanghai Municipality. National Development and Reform Commission (NDRC) Climate Change Department. Pilot measures for Shanghai's carbon emission trading mechanism [EB/OL] NDRC Climate Change Department website, http://qhs.ndrc.gov.cn/qjfzjz/201312/t20131231_697049.html.
6. Chongqing Municipality Development and Reform Commission. The notice on regulations for emission quota management of Chongqing carbon emission trading (for trial) [YuFaGaiHuan (2014) No. 538] [EB/OL]. Chongqing DRC website, http://www.cqdpc.gov.cn/article-1-20496.aspx.
7. Shenzhen Development and Reform Commission. Shenzhen had completed the annual carbon emission performance of 2013[N/EB].China carbon emission trading internet. 2014-07-05 [2016-08-15]. http://www.tanpaifang.com/tanjiaoyisuo/2014/0704/34753.html.
8. SHENZHEN Carbon Trading Task Group. On a trading system for carbon emission by macro and structural regulating[J]. China opening journal, 2013, 168(3): 7–17.
9. JIANG Jingjing. On carbon quota distribution on a limited rationality repeat game[J]. China opening journal, 2013, 168(3): 18–35.
10. People's Government of Hubei Province. Pilot measures for Hubei's carbon emission trading mechanism [EB/OL]. Hubei People's Government website, http://gkml.hubei.gov.cn/auto5472/auto5473/201404/t20140422_497476.html.

11. People's Government of Guangdong Province. The implementation proposal of Guangdong's pilot carbon emission trading mechanism [EB/OL]. 2012-09-07[2016-08-15] http://zwgk.gd.gov.cn/006939748/201209/t20120914_343489.html.
12. Guangdong Province Development and Reform Commission. Notice of the Guangdong provincial development and Reform Commission on Issuing the first allocation and work plan for carbon emission permits (for Trial Implementation) [EB/OL], 2013-11-26[2016-08-5] http://210.76.72.13:9000/pub/gdsfgw2014/zwgk/tzgg/zxtz/201311/t20131126_230325.html.
13. Guangdong Province Development and Reform Commission. 2013 annual auction announcement of carbon emissions quota. [EB/OL].2013-12-10[2016-08-15] http://210.76.72.13:9000/pub/gdsfgw2014/zwgk/tzgg/qtgg/201312/t20131210_232286.html.
14. Guangdong Province Development and Reform Commission. Guangdong's carbon emission trading market and performance status in 2013[EB/OL]. 2014-07-15[2016-08-15]. http://210.76.72.13:9000/pub/gdsfgw2014/zwgk/gzdt/gzyw/201501/t20150129_294983.html.

Chapter 4
Guangdong Carbon Emissions Status Quo and Main Characteristics

Guangdong energy consumption mix—dominated by fossil fuels—has generated high level of carbon emissions. The statistics revealed that aggregate energy consumption had been increasing year by year, and the evolution process could be split into three stages during 1995–2012. In Stage I (1995–2002), aggregate energy consumption increased with an AAGR at 6%. During this period, Guangdong economy grew at a slow pace, and consumption of coal, petroleum, and electricity achieved at an AAGR at 5, 7, and 14%, respectively. In Stage II (2003–2008), aggregate energy consumption grew with an AAGR at 12%, while GDP reached an AAGR at 14% (at 2005 constant prices), and the IVA rose with an AAGR at 20%, the consumption of coal, petroleum, and electricity presented an AAGR at 12, 9, and 22%, respectively. There was robust energy demand from industrial sectors in this stage, which led to the electricity demand over the supply. Though the total capacity of power generation expanded 52% during this period, newly added coal-fired power plants were not enough to satisfy the shortage. Thus, electricity imported from western China sharply ascended from 21.3 TW h in 2003 to 92.8 TW h in 2008. In Stage III (2009–2012), aggregate energy consumption dropped by 10% annually [1, 2]. Due to the impacts of the global financial crisis at the end of 2018 as well as domestic economic factors like the appreciation of RMB, GDP of Guangdong slowed down with an AAGR at 12%. Provincial export-oriented manufacturing sectors, which concentrated in the Pearl River Delta, were severely stricken. As a result, energy consumption in industrial sectors began to decrease. The AAGR of coal, petroleum and electricity consumption dropped to 10, 7, and 11%, respectively. Generally, Guangdong aggregate energy consumption reached 263 Mtce in 2012, accounting for 7.7% of total energy consumption in China.

This chapter will review the changes of Guangdong energy consumption patterns over 1995–2012, and analyze the characteristic of energy consumption of Guangdong province, including: carbon emission flow, emissions structure, carbon emission intensity, per capita emissions, emissions feature of electricity generation, emissions from energy consumption, and driving forces of carbon emissions.

© China Environment Publishing Group Co., Ltd. and Springer Nature Singapore Pte Ltd. 2019
D. Zhao et al., *A Brief Overview of China's ETS Pilots*,
https://doi.org/10.1007/978-981-13-1888-7_4

4.1 Guangdong Aggregate Energy Consumption Mix

Guangdong aggregate energy consumption, by maintaining an upward trend over 1995–2012 (see Fig. 4.1), reached 263 Mtce in 2012, accounting for 7.7% of China total energy consumption [2].

4.1.1 Guangdong Aggregate Energy Consumption and Spatial Distribution

The total energy consumption per unit GDP (TEC/GDP) across Guangdong maintained a downward trend over 2005–2011, i.e., the TEC/GDP of the entire province fell from 0.794 to 0.640 tce/10^4 yuan, dropping by an average of 3.53% annually. On a national scale, Beijing had the lowest TEC/GDP at 0.549 tce per 10,000 yuan in 2011, accounting for 62–45% of the national average. Ningxia had the highest TEC/GDP at 3.497 tce per 10,000 yuan in 2011. Guangdong had a fairly low TEC/GDP. Among all Chinese provinces, Guangdong was the only one that ranked second to Beijing, indicating that the energy-based economy and technologies of Guangdong were ranking at a relatively high level. However, owing to varied industrial structure and technical strength in different areas, the TEC/GDP

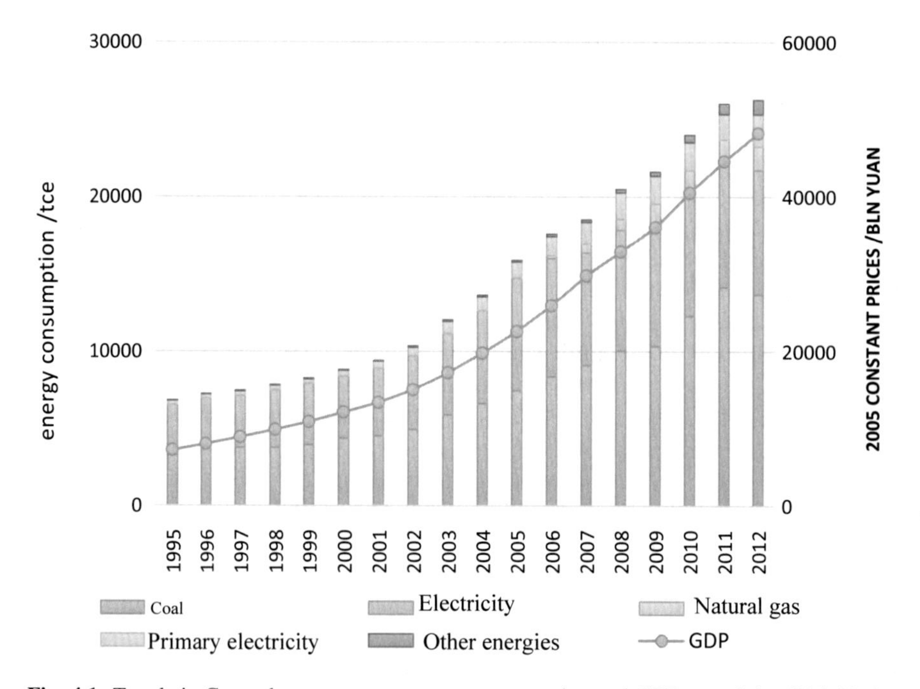

Fig. 4.1 Trends in Guangdong aggregate energy consumption and GDP growth in 1995–2012

of Guangdong varied greatly among different areas. Among the province's four major areas (see Table 4.1), the developed Pearl River Delta had energy use efficiency significantly above the other areas. The delta-based Shenzhen City had TEC/GDP no more than 0.5 tce/10^4 yuan, in contrast, Shaoguan in north Guangdong had the highest TEC/GDP, which was three times or more above Shenzhen [1, 3, 4].

With a view to TEC/GDP spatial distribution in Guangdong, in 2005, only two cities (Shenzhen and Shanwei) had a TEC/GDP lower than 0.6 tce/10,000 yuan; 13 cities were in range of 0.6–1.2 tce/10,000 yuan; 5 cities in range of 1.2–1.8 tce/10,000 yuan; only Shaoguan above 1.8 tce/10,000 yuan. In 2011, 5 cities (Shenzhen, Shanwei, Guangzhou, Zhuhai, and Shantou) had a TEC/GDP lower than 0.6 tce/10,000 yuan; 13 cities were in range of 0.6–1.2 tce/10,000 yuan; 3 cities in range of 1.2–1.8 tce/10,000 yuan; no city was above 1.8 tce/10,000 yuan.

Table 4.1 Guangdong TEC/GDP in 2005–2011 (Unit: tce per 10,000 yuan)

		2005	2006	2007	2008	2009	2010	2011
National		**1.28**	**1.24**	**1.18**	**1.12**	**1.08**	**1.03**	**1.01**
Beijing		**0.792**	**0.760**	**0.714**	**0.662**	**0.606**	**0.582**	**0.549**
Ningxia		**4.140**	**4.099**	**3.954**	**3.686**	**3.454**	**3.308**	**3.497**
Guangdong		**0.794**	**0.771**	**0.747**	**0.715**	**0.684**	**0.664**	**0.640**
Pearl River Delta	Guangzhou	0.782	0.746	0.713	0.680	0.651	0.621	0.590
	Shenzhen	0.593	0.576	0.560	0.544	0.529	0.513	0.490
	Zhuhai	0.659	0.640	0.624	0.603	0.581	0.560	0.536
	Foshan	0.888	0.848	0.811	0.746	0.694	0.664	0.637
	Dongguan	0.864	0.822	0.778	0.738	0.705	0.691	0.658
	Zhongshan	0.779	0.738	0.701	0.673	0.646	0.636	0.610
	Jiangmen	0.872	0.865	0.833	0.777	0.732	0.715	0.688
	Zhaoqing	0.988	0.958	0.919	0.887	0.844	0.823	0.793
	Huizhou	0.856	0.976	1.016	0.956	0.947	0.892	0.856
North (mountainous area)	Shaoguan	2.140	2.038	1.913	1.819	1.737	1.710	1.490
	Heyuan	0.962	0.897	0.884	0.840	0.809	0.800	0.771
	Meizhou	1.433	1.398	1.333	1.279	1.229	1.189	1.137
	Qingyuan	1.734	1.695	1.633	1.540	1.481	1.452	1.394
	Yunfu	1.525	1.436	1.389	1.338	1.294	1.274	1.228
West	Yangjiang	0.869	0.795	0.763	0.736	0.709	0.702	0.677
	Zhanjiang	0.738	0.710	0.682	0.659	0.641	0.639	0.616
	Maoming	1.332	1.282	1.256	1.190	1.146	1.097	1.055
East	Chaozhou	1.469	1.417	1.368	1.321	1.274	1.232	1.186
	Jieyang	1.029	0.973	0.940	0.902	0.874	0.855	0.819
	Shantou	0.692	0.662	0.648	0.632	0.608	0.588	0.567
	Shanwei	0.579	0.570	0.557	0.559	0.528	0.517	0.497

Note The TEC/GDP is calculated at 2005 constant prices

Source China Statistical Yearbook 2012, Guangdong Energy Statistics 2001–2010 and Guangdong Statistical Yearbook 2012

In terms of geographical distribution, the cities with a relatively high TEC/GDP are concentrated in the northern and eastern areas of Guangdong, while the cities with a lower TEC/GDP are located in the Pearl River Delta and coastal area which are fairly developed.

4.1.2 Elasticity Coefficient of Energy Consumption

Elasticity Coefficient of Energy Consumption (ECEC) is the percentage change in energy consumption to achieve one percent change in national GDP at constant prices. It is an indicator to demonstrate the relation between energy consumption and GDP growth. A smaller ECEC represents less dependence of GDP growth on energy consumption. Figure 4.2 shows that Guangdong ECEC had been fluctuating over 1995–2012 with both GDP and energy consumption maintaining positive growth in each year; thus, ECEC could remain as a positive value. There had been ECEC >1 in 2003, 2005, and 2008, showing Guangdong GDP growth was notably lower than energy consumption growth. There was ECE <1 in the most part of this period, indicating that although Guangdong kept developing its socio-economy, its energy consumption was increasing at a slower pace, but the aggregate energy consumption did not show any downward momentum [1, 2].

After comparing the ECEC between developed countries/regions and Guangdong (Table 4.2), we've found out that Guangdong had the highest growth rate in both GDP and energy consumption over 1995–2012. Guangdong had an ECEC slightly above the national average. In contrast to the US, Germany, UK, and Japan with a low or negative ECEC over 1995–2012, the ECEC of Guangdong and China at large remained high, demonstrating that the Chinese economy (including

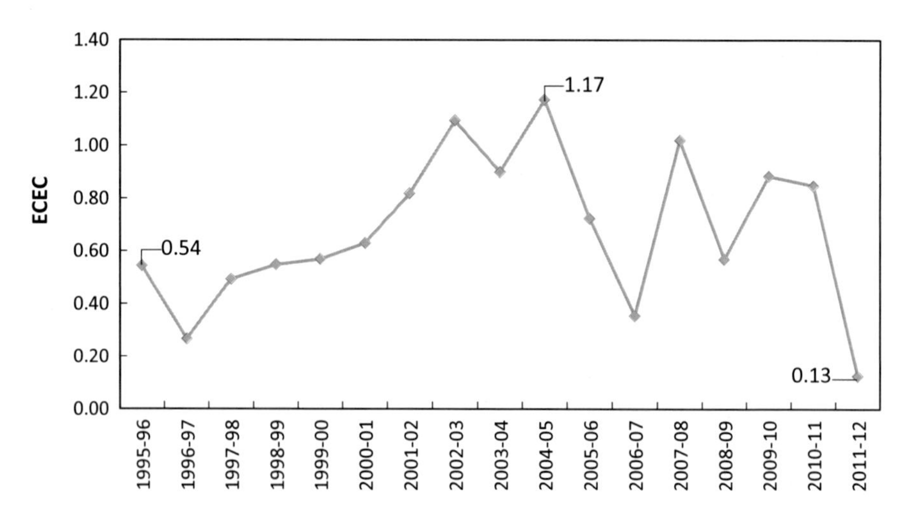

Fig. 4.2 Guangdong ECEC in 1995–2012

Table 4.2 ECEC in certain countries/regions in 1995–2012

Country/Region	GDP growth (%)	Energy consumption growth (%)	ECEC
Mainland China (1)	382	176	0.46
Guangdong (1)	571	286	0.50
US (2)	52	3	0.06
Germany (2)	26	−9	−0.34
UK (2)	43	−11	−0.25
Japan (2)	14	−9	−0.66
S. Korea (2)	111	82	0.74
Hongkong (2)	81	39	0.49

Source (1) China Energy Statistical Yearbook 1996–2013; (2) the World Bank

Guangdong) was growing at a faster pace though started later, registering a GDP growth rate significantly above the growth rate of energy consumption, hence ECEC <1. However, in contrast to Germany, UK, and Japan where the energy demand kept negative growth, Guangdong had an actively increasing energy demand with a positive ECEC.

4.1.3 Energy Consumption Intensity

Although Guangdong aggregate energy consumption kept increasing over 1995–2012, the Energy Consumption Intensity (ECI) dropped from 0.95 tce per 10,000 yuan in 1995 to 0.55 tce per 10,000 yuan in 2012 (calculated at 2005 constant prices), registering a decrease rate at 43%. In 2012, the ECI of Guangdong was lower by 41% than the national average, but still higher than that of developed countries/regions like US, Germany, UK, Japan, South Korea, and Hongkong (see Fig. 4.3). However, with a view to ECI variation trend over 1995–2012, the ECI of Guangdong and China at large, respectively, dropped by 43 and 45%, higher than such decrease between 14 and 38% in US, Germany, UK, Japan, South Korea, and Hongkong [1, 2, 5].

4.1.4 Energy Consumption Per Capita

In 2012, the GDP (at current prices), population and aggregate energy consumption of Guangdong, respectively, accounted for 11.0, 7.8 and 7.7% of the national total. Along with constantly increasing GDP per capita, the energy consumption per capita (ECpc) of Guangdong rose from 0.92 tce in 1995 to 2.48 tce in 2012, registering a robust growth rate at 169% (see Fig. 4.4). In 2012, Guangdong ECpc was 17% below the national average level, and respectively, lower by 74, 55, 43,

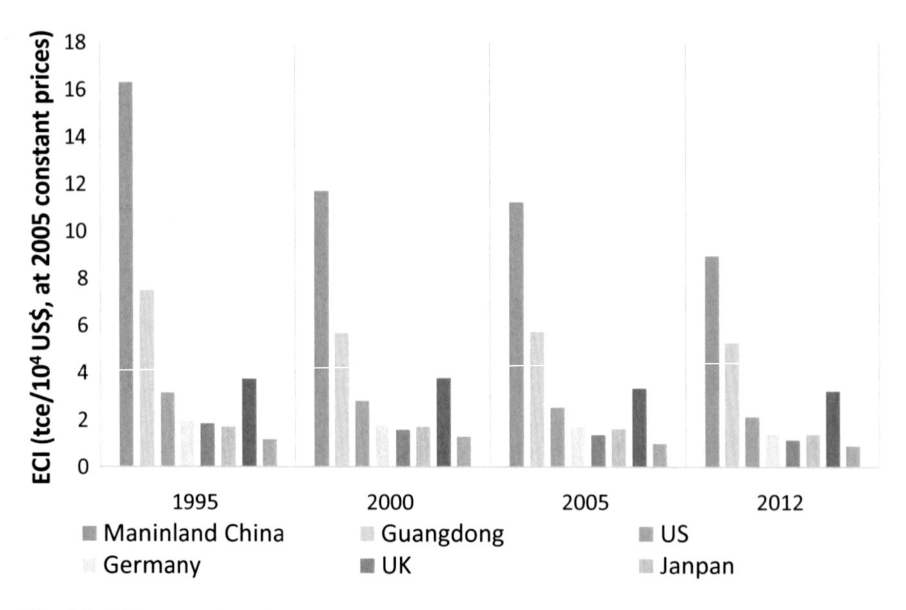

Fig. 4.3 ECI comparison in 1995–2012

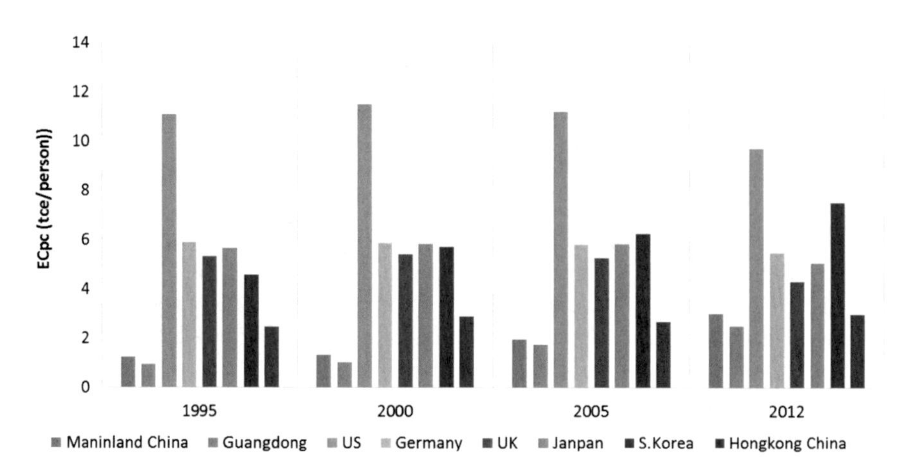

Fig. 4.4 ECpc comparison in 1995–2012

51, 67 and 16% in contrast to US, Germany, UK, Japan, South Korea, and Hongkong. However, with a view to ECpc growth rate, Guangdong ECpc growth rate was higher by 27% than the national average level over 1995–2012. In the meanwhile, the ECpc growth rate of US, Germany, UK, and Japan dropped by 12, 7, 19, and 11%, respectively. The ECpc growth rate of South Korea and Hongkong rose by 64% with 20%, respectively, yet far below that of Guangdong [1, 2, 5].

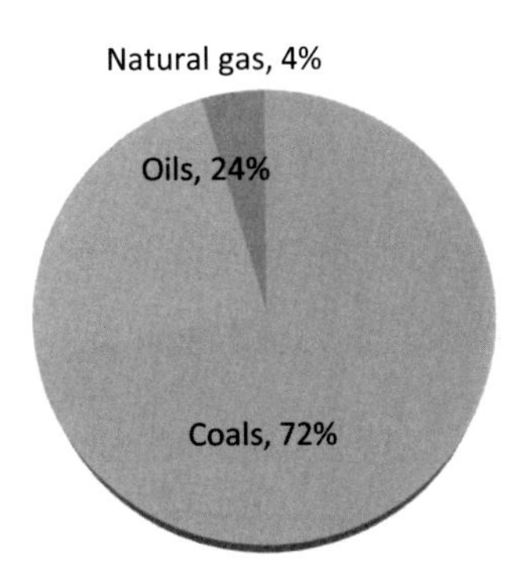

Fig. 4.5 Guangdong aggregate carbon emissions breakdown by fuel type in 2012. *Note* Guangdong aggregate carbon emissions reached 573 $MtCO_{2e}$ in 2012

4.2 Guangdong Aggregate Carbon Emissions and Carbon Flow

In light of *Guangdong Statistical Yearbook 2013*, and calculating method recommended by IEA and U.S.A, [6, 7] Guangdong aggregate carbon emissions reached 573 $MtCO_{2e}$ in 2012; among them, the emissions from consumption of coal, oil, and natural gas were, respectively, 411, 137, and 25 $MtCO_{2e}$. Their percentages in aggregate carbon emissions are shown in Fig. 4.5.

With a view to China's carbon emissions scale by province/region/municipality,[1] Guangdong was among the heavy emissions rank over 1995–2005, then joined in the ultra-heavy emissions rank since 2006. Of the emissions arising from production and livelihood, it was the emissions from the primary industry were decreasing amid moderate fluctuations, i.e., down from 1.46 tCO_{2e} in 1995 to 1.23 tCO_{2e} in 2011, registering an annual average decrease at 1.03%. The carbon emissions from the secondary and tertiary industries and from livelihood kept increasing, i.e., rising from 35.80 tCO_{2e}, 4.03 tCO_{2e} and 3.24 tCO_{2e} to 124.35 tCO_{2e}, 18.38 tCO_{2e} and 8.62 tCO_{2e}, with an AAGR at 8.09, 9.94 and 6.32%. Obviously, the secondary industry was the largest emitter in Guangdong with the percentage in aggregate emissions remaining in range of $(81 \pm 1)\%$. The tertiary industry was the second largest emitter in Guangdong with a share in aggregate emissions maintaining an upward trend (Fig. 4.6).

See the carbon flows of Guangdong in Fig. 4.7 [8], electricity production, which is based on energy conversion, was the largest emitter with carbon emissions of 264 $MtCO_{2e}$ in 2012, holding 46% of the provincial total emissions. The emissions from heat production reached 19 $MtCO_{2e}$, holding 3% of the provincial total emissions. After assigning the emissions from electricity and heat production to manufacturing sector—the largest emitter (holding 65%) at the end of final energy

[1]The Chinese carbon emissions scale are split into four classes: Class I (ultra-heavy emissions) with emissions larger than 1×10^8 t/year; Class II (heavy emissions) with emissions of $(9999–3000) \times 10^4$ t/year; Class III (general emissions) with emissions of $(2999–1000) \times 10^4$ t/year; Class IV (light emissions) with emissions at or below 999×10^4 t/year.

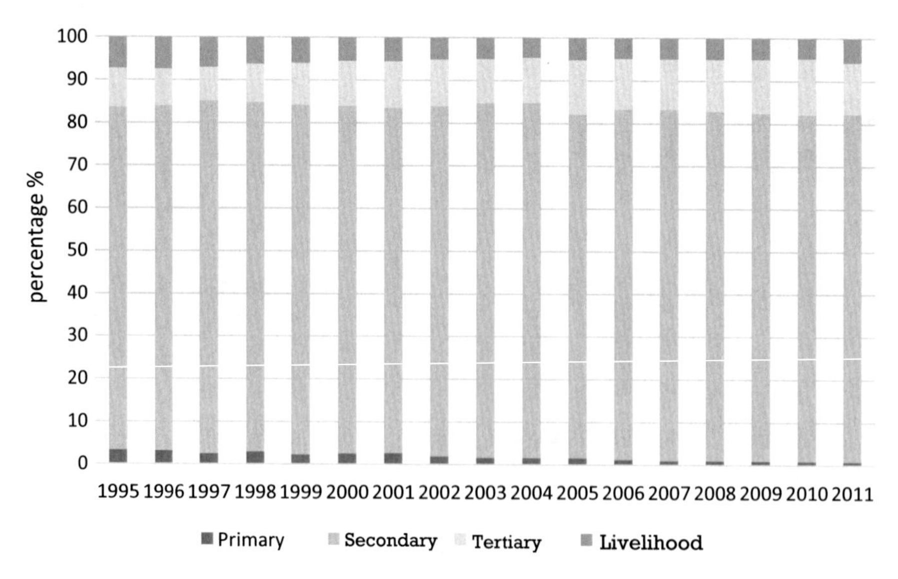

Fig. 4.6 Guangdong aggregate carbon emissions breakdown by emitting sources in 1995–2011

consumption, which was followed by transportation and storing sectors (11%); the combined emissions from the urban and rural residential consumption held 13%.

4.2.1 Characteristics of Carbon Emissions Composition

Guangdong's direct carbon emissions from all types of energy over 1995–2012 are shown in Fig. 4.8. During this period, the carbon emissions from energy consumption remained on the rise, which was attributed to mounting demand for various fossil fuels arising from the rapidly developing socio-economy. There used to be low consumption of natural gas in 1995; however, along with the operation of the receiving terminals and implementation of the "gas transmission from west China to east China", Guangdong began to consume more and more natural gas, which generated increasing carbon emissions. However, natural gas is a low-carbon energy with carbon emissions coefficient merely 56 and 72% of coal and petroleum. According to a rough estimation shows that replacing coal or petroleum consumption with natural gas is able to save carbon emissions by 44 and 28%.

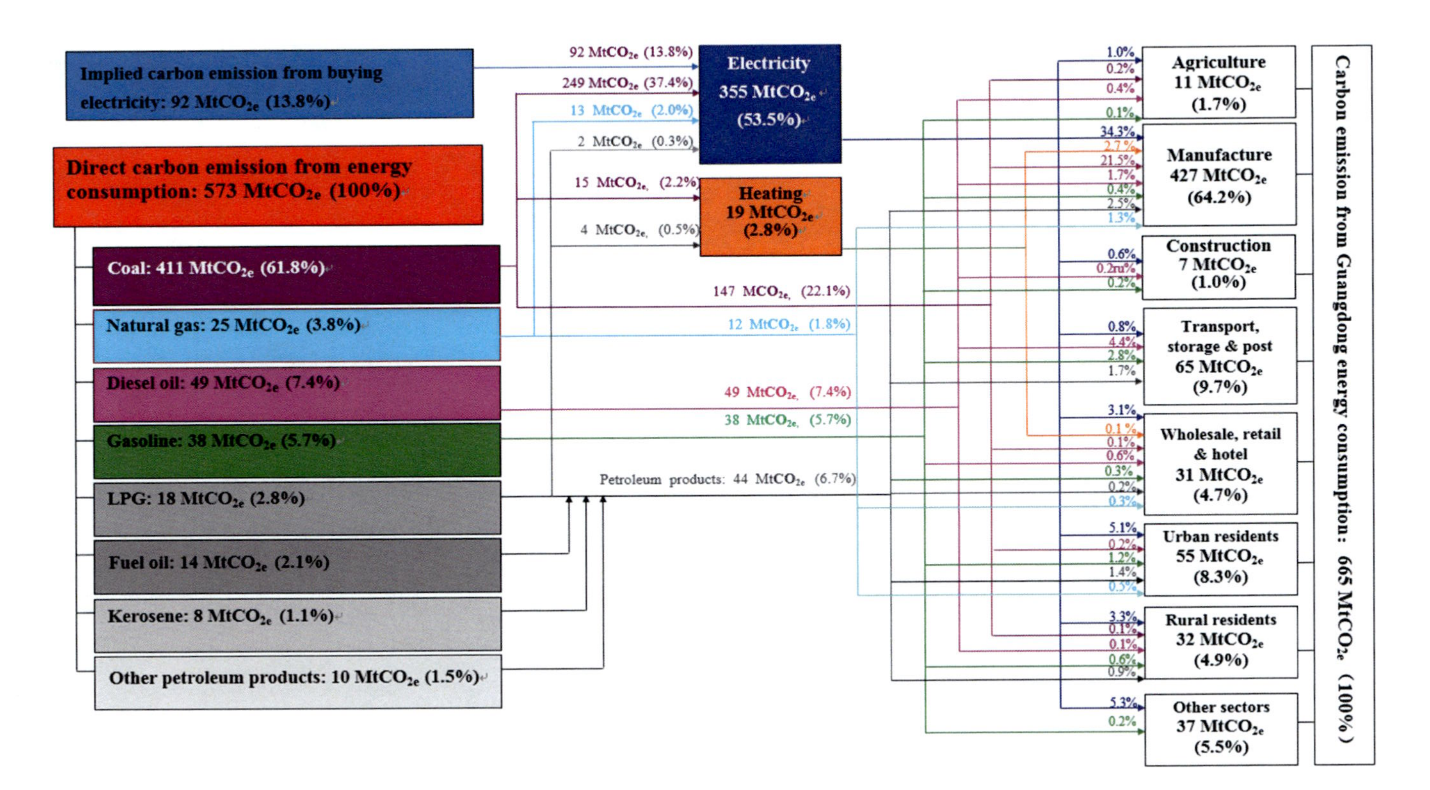

Fig. 4.7 Guangdong's carbon emission flows in 2012

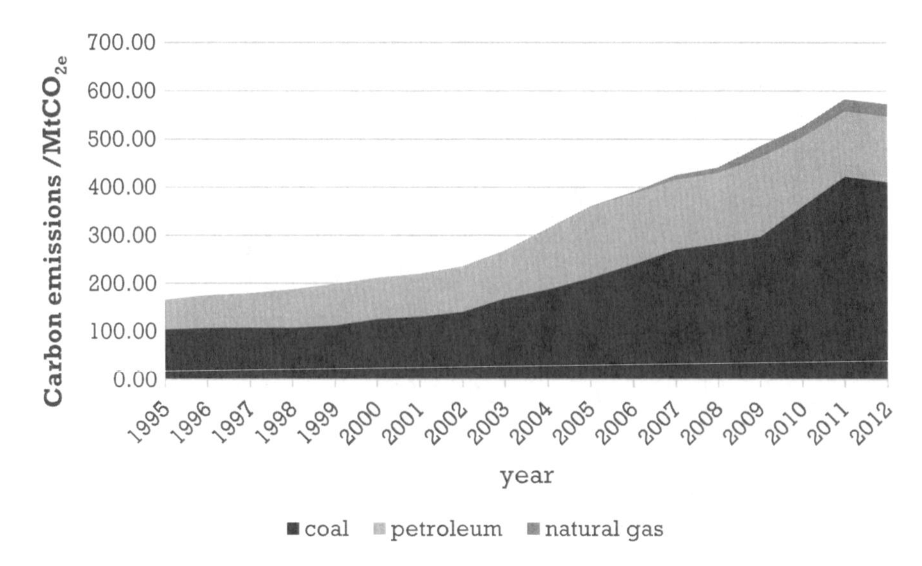

Fig. 4.8 Guangdong carbon emissions by energy type in 1995–2012

4.2.2 *Carbon Intensity*

Carbon intensity is a ratio between local aggregate carbon emissions and GDP. Quantified carbon intensity is able to reflect the dependence degree of economic growth on fossil energy consumption, it is closely related to local energy use efficiency and energy consumption mix. As shown in Fig. 4.9, both GDP and carbon emissions of Guangdong maintained an upward momentum over 1995–2012, with carbon intensity remaining at the downside, i.e., falling from

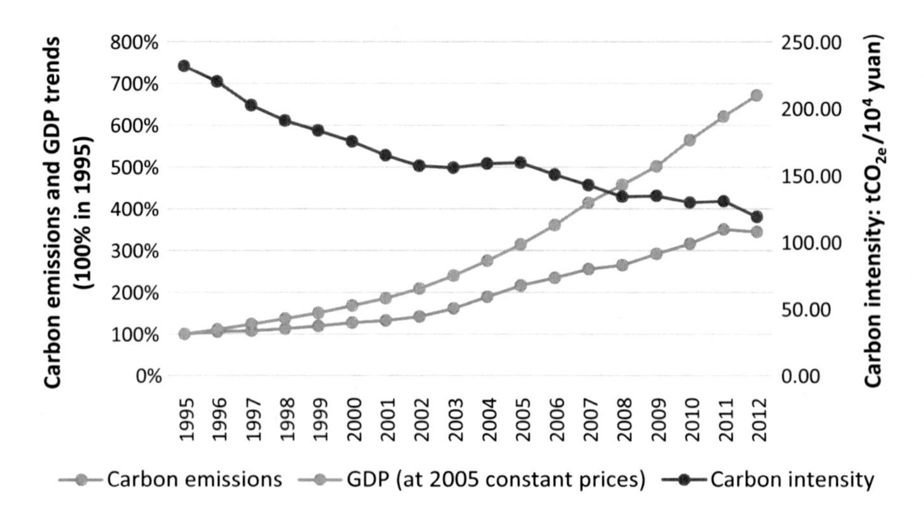

Fig. 4.9 Guangdong carbon intensity in 1995–2012

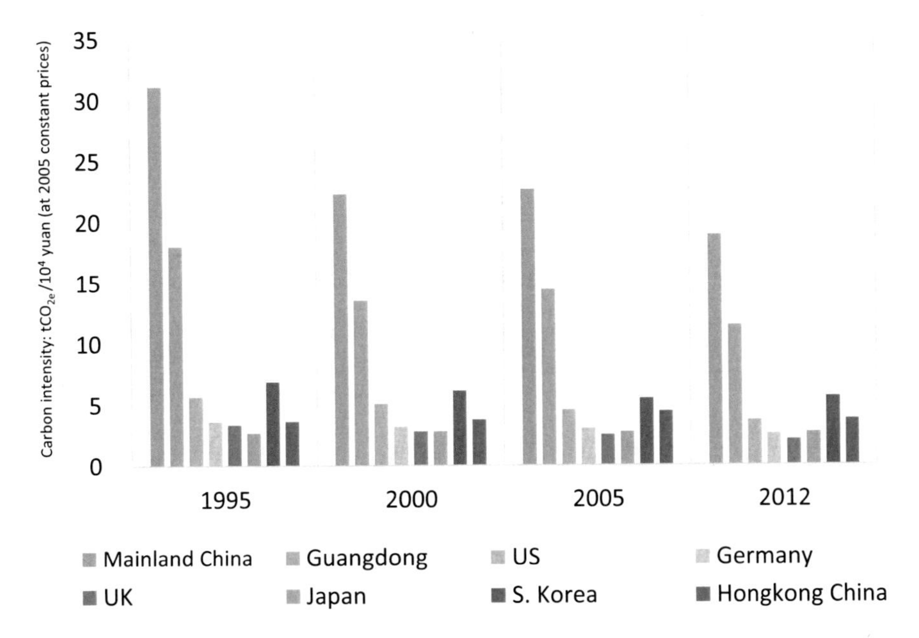

Fig. 4.10 Carbon intensity comparison between Guangdong and world major economies in. *Source* Despite the statistics about Guangdong, all the other data stem from US Energy Information Administration

2.32 tCO_{2e}/10,000 yuan in 1995 to 1.19 tCO_{2e}/10,000 yuan, dropping by an average of 4% annually.

As indicated in Fig. 4.10, Guangdong carbon intensity was lower by 65% than the national average level in 2012, but higher by 68, 78, 82, 77, 51, and 67% [5, 9] than US, Germany, UK, Japan, South Korea, and Hongkong, respectively. But with a view to the downward trend of carbon intensity, both Guangdong and China at large reduced carbon intensity by 37 and 39%, slightly higher than such drop in US (35%), Germany (30%), and UK (38%), and far above Japan (1%), and South Korea (18%), marking that Guangdong has achieved significant progress in raising energy use efficiency, energy conservation, and emissions reduction, and in upgrading socioeconomic development quality, yet there is still a long way to go for Guangdong to catch up with the advanced energy use efficiency in the European states and US.

4.2.3 *Carbon Emissions Per Capita*

Guangdong saw its population increase by 43% over 1995–2012, and reach 106 million in 2012. As shown in Fig. 4.11 the growth of Guangdong carbon emissions exceeded that of population, which resulted in a year-on-year increase in carbon emissions per capita (CEPC). In 2012, Guangdong CEPC reached 5.41 tCO_{2e}/person.

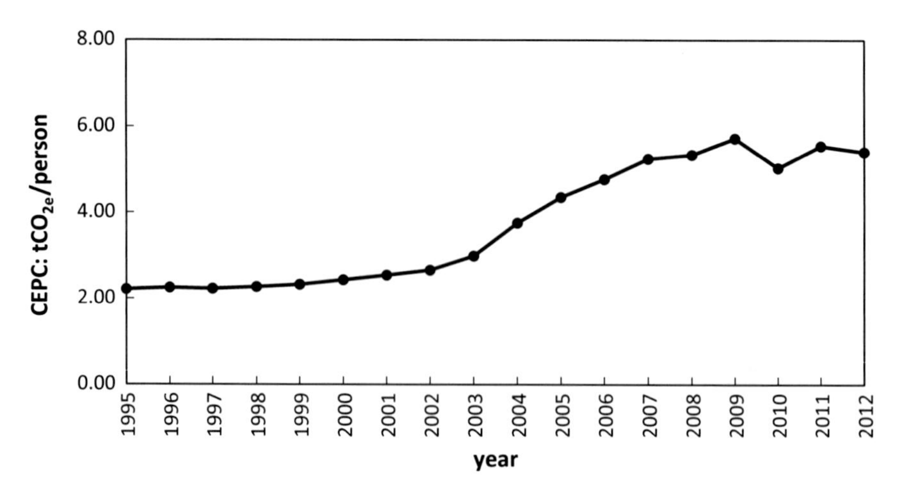

Fig. 4.11 Guangdong CEPC in 1995–2012

In 2012, the CEPC of Guangdong was 17% lower than the national average level, and remarkably lower than that of world major economies like US, Germany, UK, Japan, South Korea, and Hongkong. Over 1995–2012, the CEPC of Guangdong and China at large kept increasing, but the CEPC of US, Germany, and UK remained on the downside, falling by 16, 10, and 19%, respectively. The CEPC of Japan, South Korea, and Hongkong, though situated in Asia as Guangdong, was increasing slowly with a growth rate of 11, 55, and 61% [5, 9], respectively. Though the CEPC of Guangdong is currently lower than that of developed states/regions in Europe and America, it has been increasing rapidly; moreover, the cardinal number of Guangdong population is large and remaining stable, and the provincial economy keeps sustainable growth, which makes control of carbon intensity an especially important task (Fig. 4.12).

4.2.4 Carbon Emissions from Electricity Production

In 2012, the total carbon emissions from electricity production of Guangdong reached 264 $MtCO_{2e}$, not counting into the indirect emissions from consumption of purchased electricity. Of the total emissions, the coal-based, natural gas-based, and petroleum-based electricity production accounted for 94.4, 4.9, and 0.6%, respectively. As shown in Fig. 4.13, total carbon emissions from electricity production were assigned to each consumption end based on their percentage of total electricity consumption, so as to identify the emissions from end users. Industrial sectors—the largest emitter—contributed to 64% of total carbon emissions from electricity production; while the combined emissions from urban and rural household consumption shared another 16%.

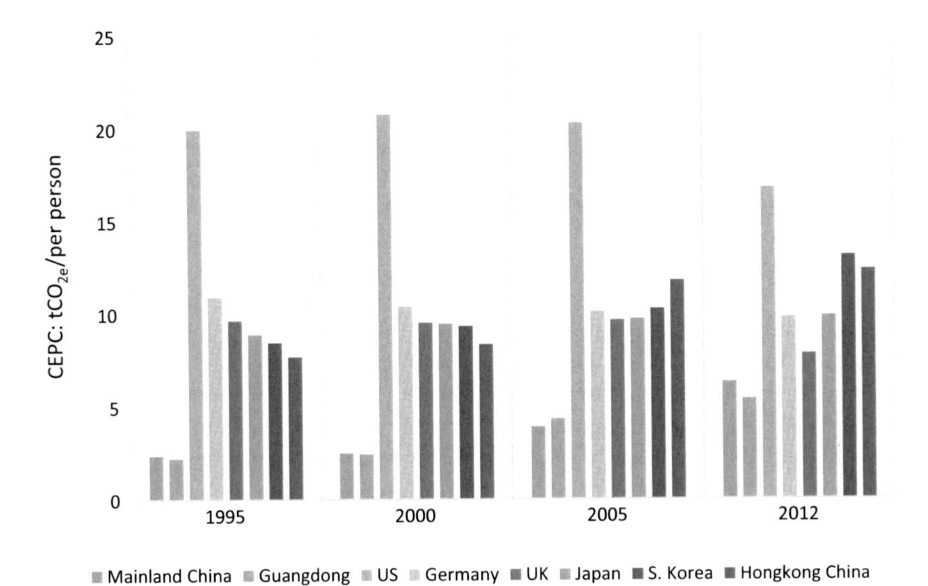

Fig. 4.12 CEPC comparison between Guangdong and world major economies in 1995–2012

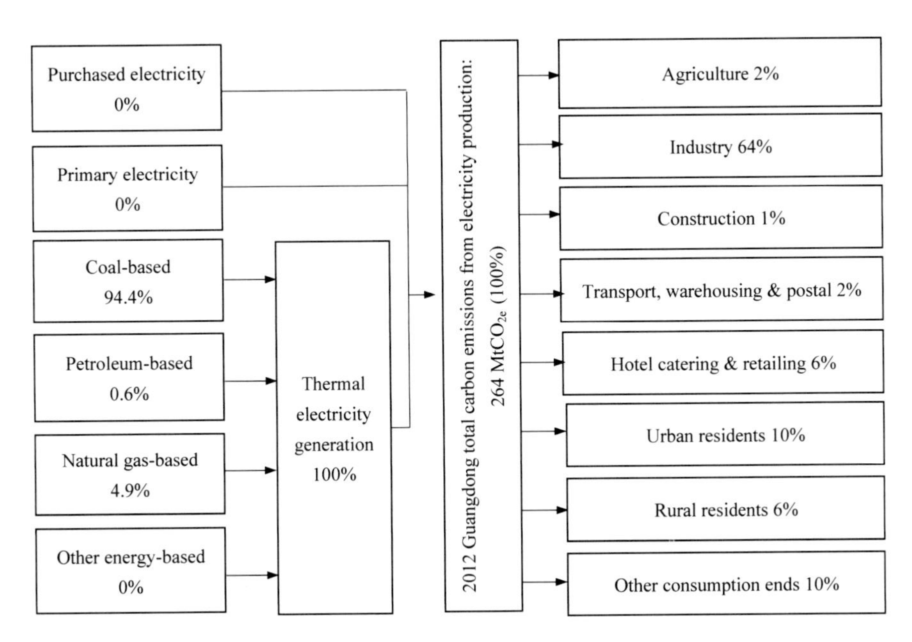

Fig. 4.13 Carbon flow comes from electricity production in Guangdong in 2012. *Note* The "primary electricity" is made up of the nuclear power, hydropower, and wind power produced within Guangdong

The characteristics related to energy consumption and carbon emissions from electricity generation of Guangdong could be analyzed by three stages, which are as follows:

Stage I (1995–2002): During this period, the economy of Guangdong was developing at a slow pace, with the energy consumption for electricity production rising at an AAGR of 14%. Of the total electricity, respective 60 and 27% were generated from coal-fired power plants and petroleum-fired power plants, 2% was from purchased electricity, and 11% was supplied by nuclear power, hydropower, and wind power together. The carbon emissions from Guangdong electricity production rose with an AAGR at 9%.
Stage II (2003–2008): During this period, the GDP of Guangdong was rising with an AAGR of 14% (calculated at 2005 constant prices), such growth of IVA was 20%, and the energy consumption for electricity production was increasing with an even higher AAGR of 22%. The coal-based installed capacity increased 52%. The share of coal-based electricity in total electricity output rose to 66%, the share of petroleum-based electricity dropped to 14% under the pressure from mounting generation cost and strict environmental protection, while the combined share of nuclear power, hydropower and wind power was 9%. In order to make up for the shortage in electricity demand, Guangdong increased electricity purchase from the project as "electricity transmission from west China to Guangdong", i.e., from 21.3 TWh in 2003 to 92.8 TWh in 2008. The share of purchased electricity in Guangdong total electricity output rose to 13%. Owing to the sharp increase in energy consumption for electricity production, and shrinking share of nuclear power, hydropower, and wind power, the carbon emissions from Guangdong electricity production rose with an AAGR at 10% during this period.
Stage III (2009–2012): Under the impact from several economic factors at the end of 1998 like the global financial storm and appreciation of RMB, the GDP of Guangdong was increasing with a lower AAGR at 10% during this period. Such growth of the electricity consumption in industrial sectors dropped from 13% over 2003–2008 to 7%. Such growth of the energy consumption for electricity production fell to 7%. Of the total electricity output, the share of electricity from coal-fired generators rose to 72%. Owing to sharply increasing generation cost of petroleum-fired power plants and intensified efforts in air pollution control, small thermal power plants with a capacity of about 12.1 GW were shut down during 2006–2010, the share of petroleum-based electricity dropped to 1%. Along with Shenzhen Dapeng LNG Receiving Terminal stating to operate in 2006, the share of electricity from natural gas-fired power plants achieved to 7%. The amount of electricity purchased from other provinces based on the program of "electricity transmission from west China to Guangdong" steadily ascended, and its share in total electricity supply increased to 11% consequently. Meanwhile, electricity supplied by nuclear power, hydropower, and wind power together took account for rest 9%. Overall, the carbon emissions from power sectors increased with an AAGR of 8% in this stage (Fig. 4.14).

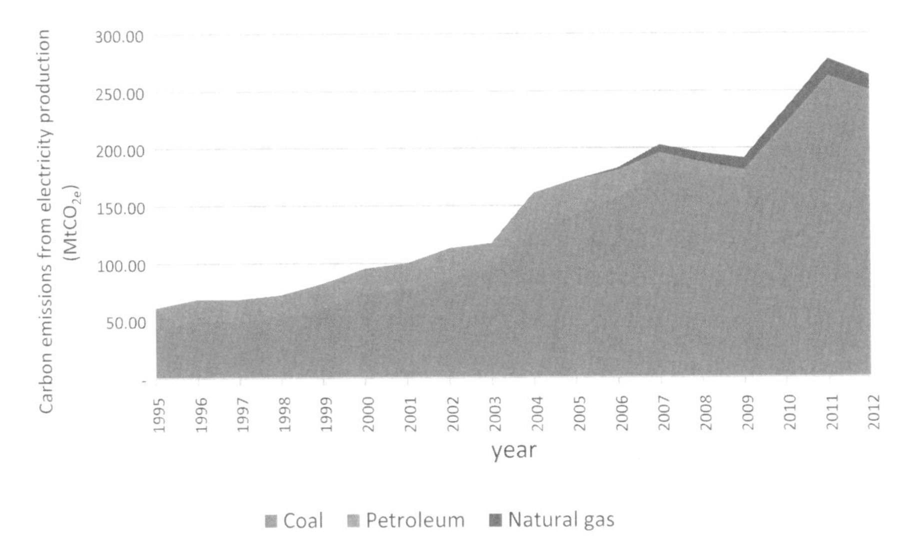

Fig. 4.14 Carbon emissions from Guangdong electricity production by different energies in 1995–2012

4.3 Decomposition of the Driving Force for Carbon Emissions

In order to analyze the characteristic carbon emissions from end users, emissions from electricity and heat generation were then allocated to respective end users based on energy consumption percentage. Overall, the aggregate carbon emissions of Guangdong from energy consumption rose from 166 $MtCO_{2e}$ in 1995 to 573 $MtCO_{2e}$ in 2012, and emissions from all end users had kept an upward trend (see Fig. 4.15).

In the percentage of emissions, manufacturing sector was ranking the top (64–68%) over 1995–2012. During this period, the percentage of emissions from transport, warehousing and postal sectors, wholesale-retail sectors, and other service sectors rose 3, 1, and 2%, respectively; such percentage of agriculture, animal husbandry, and fishery sectors dropped 2%.

With a view to the growth rate of emissions, the AAGR of the emissions from transport, warehousing and postal sectors, wholesale-retail sectors, and other service sectors was 10%, higher than that of the provincial total emissions (8%). The manufacturing sector, which belongs to the secondary industry, is the largest emitting source. The AAGR of carbon emissions from manufacturing sector leveled off the growth of provincial total emissions. In contrast, such growth of agriculture, animal husbandry and fishery sectors, which belong to the primary industry, was merely 2%. Such situation is mainly attributed to the industry restructuring in Guangdong, i.e., the focus is shifting from the primary industry to tertiary industry,

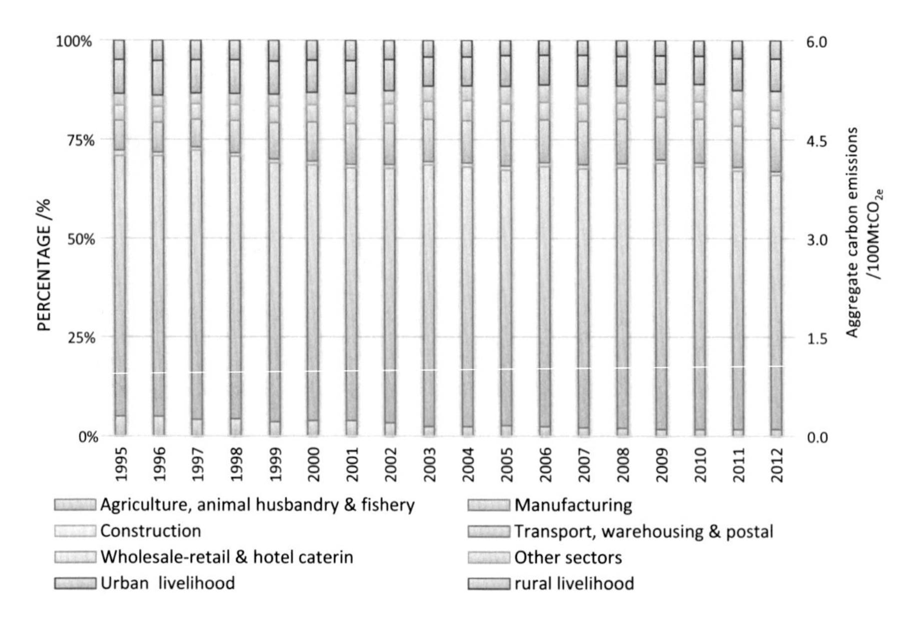

Fig. 4.15 Percentage of carbon emissions of Guangdong by consumption end in 1995–2012

yet the adjustment to the secondary industry is limited; in other words, the secondary industry remains as the predominant emitting source.

With a view to the increments in carbon emissions, the aggregate carbon emissions of Guangdong increased 407 $MtCO_{2e}$ over 1995–2012. Of the total increments, 64% was contributed by the manufacturing sector, 12% were from transport, warehousing, and postal sectors, and 13% from urban and rural livelihood.

In order to dissect the driving forces for carbon emissions from energy consumption end, we use LMDI approach to interpret the Kaya and Johan models that represent development of industrial sectors, and the ImPACT model indicating development of population and society, so as to analyze the 10 driving forces which are distributed in production sector (i.e., economic activities, industry structure, energy intensity, energy consumption mix, and emissions coefficient of energy consumption) and in people's livelihood (i.e., population growth, urbanization level, degree of prosperity, energy consumption mix, and emissions coefficient of energy consumption).

Of the total the increments in carbon emissions of Guangdong over 1995–2012, 64% was contributed by manufacturing sectors; 2% by "transport, warehousing and postal sectors"; 8 and 5% by urban and rural livelihood, respectively; 6, 5, and 1% by the other service sectors, wholesale-retail and hotel catering sectors and construction sector, respectively (see Fig. 4.16).

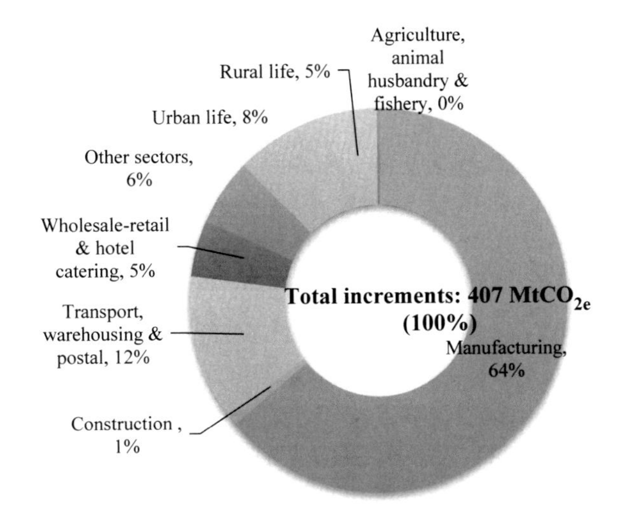

Fig. 4.16 Breakdown of Guangdong carbon emissions increments by energy end-use in 1995–2012

4.3.1 Driving Force in Carbon Emissions from Production End

The increments in carbon emissions from production end[2] reached 356 $MtCO_{2e}$ over 1995–2012, accounting for 87% of the total increments. The carbon emissions from production end were increasing at a varied pace over 1995–2012 (see Fig. 4.17). The increments over 2003–2009 were ever larger and averaging at 32 $MtCO_{2e}$ per year, which was 55% higher than the average increments over 1995–2012. It was attributed to the robust development of the export-oriented economy of Guangdong since China's accession to the WTO, the production scale of all economic sectors began to expand, and the carbon emissions from energy consumption were increasing at a fast pace, such increments topped at 42 $MtCO_{2e}$ over 2008–2009.

With a view to driving force, it is the economic output that mainly drives up the growth of carbon emissions from production end. Despite of the 2011–2012 period, the driving force had exerted a positive effect and contributed 152% of the increments over 1995–2012. Energy intensity is a major driving force for cutting back the increments in emissions (cutting 42% over 1995–2012). Energy structure, industry structure, and emissions factor of the production end were of moderate effect on carbon emissions over 1995–2012, and fairly fluctuating; they contributed 6, 4 and −11% to the increments in carbon emissions over 1995–2012.

[2]"Production end" refers to agriculture, manufacturing and catering sectors despite of the energy consumption for people's livelihood. The "consumption end" shown in the previous subsector is made up of both production sectors and people's livelihood. In this subsection, "production end" and "resident end" are cited to differentiate all production sectors and people's livelihood.

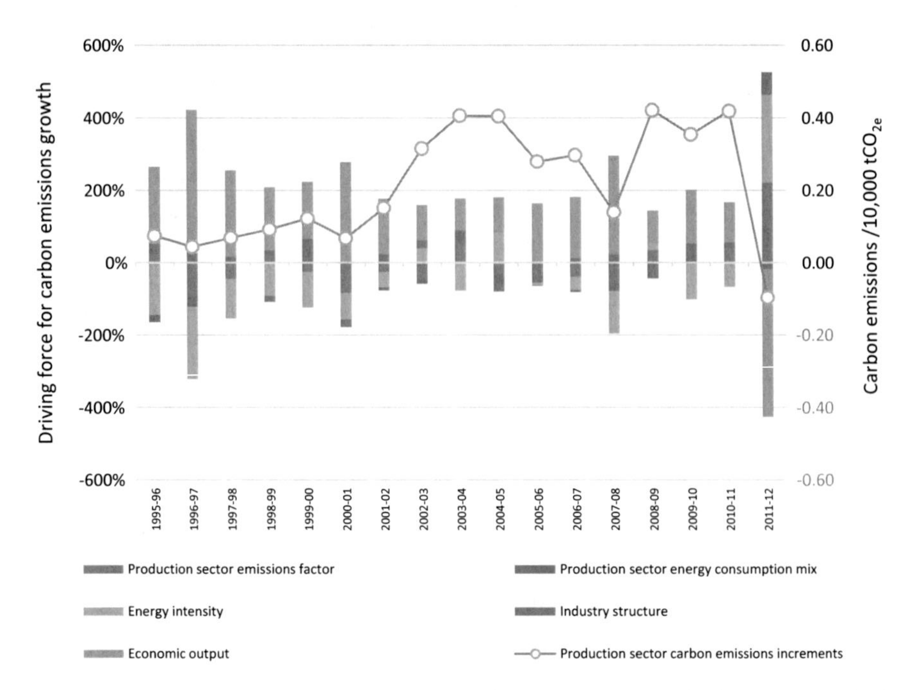

Fig. 4.17 Driving force for Guangdong carbon emissions growth in production end over 1995–2012

4.3.2 Driving Force in Carbon Emissions from Resident End

Over 1995–2012, the increments in carbon emissions from resident end reached 51 $MtCO_{2e}$, accounting for 13% of the total increments and registering a year-on-year increase. The annual average increments were as high as 5 $MtCO_{2e}$ over 2003–2011, far exceeding the increments of 1 $MtCO_{2e}$ over 1995–2002, because along with constantly developing economy and increasingly affluent life, the residents have more disposal income, and widely applied transport means and home appliances, which, in turn, increased carbon emissions. The increments topped at 15 $MtCO_{2e}$ over 2010–2011 (see Fig. 4.18).

Over 1995–2012, despite the expanding population size that exerted a positive role in driving up carbon emissions, all the other driving forces were fluctuating greatly. Overall, people's affluent life was the major driving force, contributing 76% of the increments over 1995–2012, followed by population size (26%), livelihood energy consumption mix (5%), and urbanization (4%). In addition, emissions factor—owing to optimized energy structure—slowed down increments in carbon emissions by 13%.

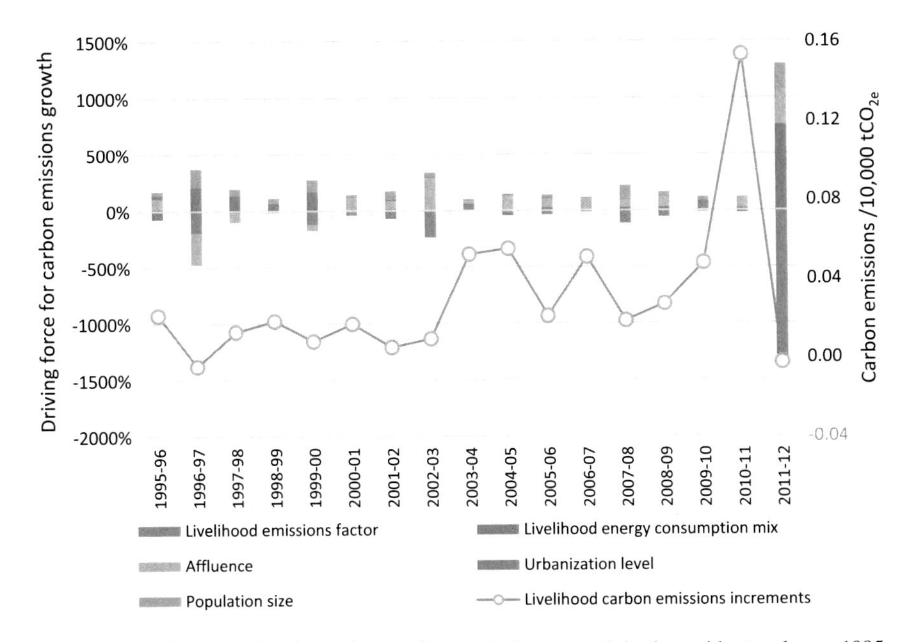

Fig. 4.18 Driving force for Guangdong carbon emissions growth in the resident end over 1995–2012

References

1. Statistics Bureau of Guangdong Province. Guangdong Statistical Yearbook 2013 [M]. Beijing: China Statistics Press, 2013.
2. NBS Department of Energy Statistics. China Energy Statistical Yearbook 2013 [M]. Beijing: China Statistics Press, 2013.
3. National Bureau of Statistics of the People's Republic of China. China Statistical Yearbook 2012 [M]. Beijing: China Statistics Press, 2012.
4. Guangdong Statistical information network. [EB/OL] http://www.gdstats.gov.cn/tjzl/tjfx/.
5. THE WORLD BANK. Indicators and Data [EB/OL]. (2014) [2014-03-31]. http://data.worldbank.org/indicator.
6. QUADRELLI R, KENNEDY A, BURGHGRAEVE S. CO_2 Emissions from Fuel Combustion: 2013 [R]. Paris: International Energy Agency, 2013.
7. U.S. ENERGY INFORMATION ADMINISTRATION. International Energy Statistics - EIA [EB/OL]. (2013) [2014-10-20]. http://www.eia.gov/cfapps/ipdbproject/IEDIndex3.cfm?tid=90&pid=44&aid=8.
8. LAWRENCE LIVERMORE NATIONAL LABORATORY. U.S. energy flow and carbon flow [EB/OL]. (2013–12) [2013-12-31]. https://flowcharts.llnl.gov/index.html.
9. INTERNATIONAL ENERGY AGENCY. IEA Sankey Diagram [EB/OL]. (2014)[2014-10-03]. http://www.iea.org/Sankey/index.html.

Chapter 5
Guangdong Pilot ETS Coverage Scope and Allowances

The ETS is comprised of seven components: coverage scope, allowances setting, allowance allocation, transaction management, legal framework, MRV rules, and linkage with external carbon markets. The coverage scope is the primary part, since its reasonable coverage, to a certain extent, decides the operational efficiency of the ETS. However, most of the Chinese current researches focus on total allowances calculation and allocation, instead of on industry coverage of the ETS. In fact, the number and ownership of the covered companies decide the allowances total amount and allocation standard of the ETS. Therefore, the research on the coverage scope shall be the fundamental work. The coverage scope of regional ETS is based on categories of GHGs and the covered industry selection. In light of the *IPCC Fourth Assessment Report: Climate Change 2007*, CO_2 is the predominant anthropogenic GHG, and also the main target in China's emissions reduction program [1]. This chapter elaborates on how to define the coverage scope appropriately for constructing ETS.

5.1 Overview the Coverage of Global ETS

The EU ETS is the world's largest GHG cap-and-trade system. It has developed all-round criteria for guiding industry selection. The EU ETS established two guiding principles for selecting the covered enterprises: First, making a narrow the coverage scope at the infancy of the ETS. Second, take the larger emitter into ETS [2]. The EU ETS had set the following five criteria for industry selection: (i) Environmental effectiveness; (ii) Economic efficiency; (iii) Impact on industrial competitiveness; (iv) Management feasibility; and (v) Substitution policy. Based on these criteria, such industrial sectors as electricity, steel, petroleum refining, glass (incl. ceramics and cement), and paper, with annual CO_2 emissions accounting for 42.6% of the total emissions of EU, were designated into the EU ETS Phase-1 regulation.

© China Environment Publishing Group Co., Ltd. and Springer Nature Singapore Pte Ltd. 2019
D. Zhao et al., *A Brief Overview of China's ETS Pilots*,
https://doi.org/10.1007/978-981-13-1888-7_5

The coverage scope of the Tokyo-ETS shall satisfy the following criteria: (i) All the buildings and installations within the administrative region of Tokyo are subject to CO_2 emissions regulation; (ii) All the buildings and installations within the administrative region of Tokyo, with annual consumption of fuels, heat, and electricity above 1500 kLOe, are involved in the ETS. The statistics show that the annual regulated CO_2 emissions account for 20% of Tokyo's total emissions [2].

The German ETS defines its coverage scope based on category and annual emissions of installations [3]. Through surveys of installations and their emissions of all companies, German Emissions Trading Authority defined that the installations (above 20 MW) as primary emitters, and they are obliged to set a cap on emissions and join in allowances trading as provided by law.

Whether the above global ETS experiences can be directly borrowed by China for building a nationwide ETS shall be first undergoing applicability analysis. A summary of the world prevailing ETS reveals that an ETS coverage scope usually satisfies the following criteria [4, 5]:

(1) Annual CO_2 emissions

The purpose for building an ETS is to fulfill regional emissions reduction goals at a lower cost. The existing ETS in all countries unanimously takes the amount of carbon emissions as a key criterion for defining regulation industries. Since "a small ship makes a good turn," the foreign countries start regulating the emitters with annual emissions taking up a small share of the regional total emissions, i.e., as small as 20–30% during Phase 1 regulation. Similarly, China shall also take account of such percentage while selecting regulation industries for the ETS. In reference to the foreign ETS experiences and the corresponding cost, while the ETS is still at its infancy, it is wise to control the percentage of regulated emissions in the country/region's total emissions at no more than 40%.

(2) Carbon leakage rate

In order to mitigate the impact of carbon trading on the domestic economy and carbon market, each country, when defining the ETS coverage scope, shall take account of the issue of carbon leakage.[1] The Kyoto Protocol targets at regulating emissions from the building; the EU ETS focuses on electric and industrial installations; the German ETS directs at industrial installations. The ETS designers at all levels shall pay special attention to carbon leakage, since it is much easier for inter-regional industry transfer. Some companies under the ETS framework may remove their high-emissions industrial production to other non-compliant areas, which may discourage the economic and industrial development of the areas of origin, and fail to fulfill the target of restricting carbon emissions. According to

[1]Carbon leakage under the Kyoto Protocol occurs when certain compliant countries transfer high-carbon industries to other noncompliant countries, thus increasing CO_2 emissions in the latter. Yet the "carbon leakage" herein occurs within one country, i.e., an area or province with a strict emissions regulation policy removes the local high-carbon industries to other areas, which may affect the economic development of the original area.

China's "12th Five-Year Plan" (2011–2015), all provinces and cities shall reduce carbon emissions, yet their reduction targets varied. Different regions were either strict or relaxed at the control of carbon emissions, which created objective conditions for carbon leakage. Therefore, the areas that set higher emissions reduction targets shall pay special attention to this issue.

(3) Management feasibility

Among the five criteria for selecting industries for the ETS regulation, "management feasibility" is especially important, since it relates to manageability of industry nature, feasibility of monitoring measures, and acceptability of management cost. These factors vary greatly among the Chinese industries, because of weak data foundation of carbon emissions and imperfect legal system about CO_2 emission. Therefore, management feasibility is a particularly important criterion for China to select ETS regulation industries.

All the ETS regulation industries in the EU, German, and Japan satisfy one common criterion: the high-emissions sources are covered to fulfill national/regional emissions reduction targets, lower average management cost, which meets the requirements for the ETS trial operation. Therefore, when selecting regulation industries for the ETS, the priority shall be given to the high-emissions sources.

5.2 ETS Coverage Selection Model Building

While designing a coverage selection mechanism of ETS, those elements of the national/regional carbon emissions structure, economic development plans, and emissions reduction targets shall all be taken into account.

5.2.1 Principles of Model Constructing

The following principles shall be followed for building an indicator system for industry selection:

(1) Integration of universality and comparability

The modules and elements in the indicator system shall be understood and accepted by all stakeholders, and able to reflect the ETS characteristics and regional differences, i.e., the selected indicators shall be of versatility and compatibility. The comparable indicators shall be selected to ensure both space–time comparison and vertical–longitudinal comparison.

(2) Scientificity and operability

When building the indicators for both element layer and base layer, the representation indicators that are scientific and able to reflect the ETS connotation and industry requirements shall be selected. Moreover, the indicators shall be easier to be collected, with authoritative and reliable data sources.

(3) Completeness and independence

The cap on regional total emission allowances trading, allowances allocation approach, and MRV design are essential contents of the ETS, thus calling for complete industry information. The indicators that are capable of satisfying the three requirements shall be selected. Besides, each indicator shall reflect unique and significant information, rather than different indicators reflecting the same fact, so as to ensure a small number of indicators reflecting the overall information of the ETS.

5.2.2 Model Construction

In light of the principles for selecting indicators, Guangdong Province designed an indicator system for the "regional ETS coverage Selection", which consists of the indicators at four levels of system layer, module layer, elements layer, and base layer for identifying the industries accessible to the ETS. This indicator system involves one Tier-1 indicator, 4 Tier-2 indicators, 10 Tier-3 indicators, and 26 Tier-4 indicators. The former three indicators are synthetic and indirect indicators, the Tier-4 indicators are objective, direct, and measurable indicators that should occupy a basic position in the indicator system.

(1) Build an indicator system for the regional ETS coverage scope

Tier-1 indicator, the only one indicator at system layer, refers to the regional industries joining in the ETS. It is a comprehensive and systematic indicator for defining the scope of ETS-regulated industries. Tier-2 indicators, which are in the module layer (below system layer), consist of four representation indicators, i.e., physical quantity, technicality, economy, and quality. Tier-3 indicators, which are an element layer, consist of 10 elements as industrial carbon emissions, emissions reduction space, emissions reduction cost, the operability of emissions administration, industrial development prospect, carbon leakage rate, trading willing, relevant policies, technical advancement, and economic efficiency. Tier-4 indicators, which are in a base layer, are made up of 26 directly measurable indicators, which mark the operability of the indicators for selecting the ETS-regulated industries.

(i) Preconditions:

Manageability of carbon emissions data is precondition or threshold for industries to join in carbon trading. It is the industries with C_3 score above 0.65 that are covered in the indicator system (this precondition is numbered as B_2, since it essentially attached to B_2).

(ii) Module of physical characterization

This module is made up of two elements: emissions amount and possible carbon leakage. The coverage of large emitters in the ETS help fulfill regional emissions reduction targets, and activate carbon trading market. Carbon leakage concerns the sustainable development of regional economy. If there was a cap on total emissions of an industry, which is fairly mobile and less dependent on resources, there would be capital flight, thus exerting a negative impact on local economy and employment, and hurt stability of the ETS operation.

(iii) Module of technical advancement

"Emissions reduction space" measures the possibility of emissions reduction through technical improvement, so as to manifest the emissions reduction potentials of the industry. "Technical advancement" represents the overall technical conditions of an industry, judge the position of the industry in the carbon market, i.e., buyer or vendor.

(iv) Module of economy

"Emissions reduction cost" measures the expenses spent for cutting emissions. Economically, a company will not adopt any costly technology even though it is able to cut CO_2 emissions. Such potential for emissions reduction is theoretical, instead of operable. "Industry prosperity" is able to differentiate sunset industries from emerging industries.

(v) Module of quality advancement

Such indicators are used to express and deal with the fairly important information that is hard to be conveyed with visual data. "Trading will" and "relevant policies", mainly through interpretation of regional emissions reduction policies and basic industrial conditions, are used to judge the attitude of stakeholders, and the obstacles for the industries to join in carbon trading.

(vi) Explanations of indicators

There are three categories of indicators in Table 5.1, quantitative indicators, nominal indicators, and sequencing indicators. D_7, D_8, D_{23}, D_{24}, and D_{25} are nominal indicators; their value only differs in categorical attributes, and means equal or unequal; if the value of D_7, D_8, $D_{24,}$ and D_{25} is set at "1 or 0" which, respectively, represents "Yes or None" or "Concentrated or Scattered". The value of D_{23} is −1 (very high: fully competitive market), −0.6 (high: monopolistic competition market), −0.3 (relatively low: oligopoly market), and −0 (low: administrative regulation).

Table 5.1 Comprehensive assessment indicators for ETS-regulated industries

Tier-1	Tier-2	Weight	Tier-3	Weight	Tier-4 Name	Tier-4 Weight	Total weight
Precondition P		***1***	***C_3 Manageability***	***1***	***D_7 Emissions monitoring and verification***	***0.40***	***0.40***
					D_8 Distribution of emissions sources	**0.60**	**0.60**
Industry accessibility A	Physical B_1	0.25	C_1 Carbon emissions	0.7	D_1 Amount of emissions	0.48	0.084
					D_2 Percentage of emissions from energy and technical process	0.21	0.037
					D_3 Percentage of regulated emissions in regional total emissions	0.21	0.037
					D_4 Unit carbon intensity	0.10	0.018
			C_2 Carbon leakage	0.3	D_5 Resource dependence	0.42	0.032
					D_6 Fixed-asset lock-in rate	0.58	0.043
	Technical B_2	0.25	C_4 Emissions reduction space	0.6	D_9 Emissions reduction potentials in production process	0.41	0.062
					D_{10} Potentials in energy consumption	0.41	0.062
					D_{11} Ranking of industrial emissions reduction space in the region	0.18	0.027
			C_5 Technical advancement	0.4	D_{12} Number of companies using low-carbon technologies	0.46	0.046
					D_{13} Maximum difference value of unit carbon intensity inside the industry	0.19	0.019
					D_{14} Ranking of industrial average carbon intensity among the national same industries	0.35	0.035

(continued)

Table 5.1 (continued)

Tier-1	Tier-2	Weight	Tier-3	Weight	Tier-4 Name	Tier-4 Weight	Total weight
	Economic B_3	0.25	C_6 Emissions reduction cost	0.4	D_{15} Average emissions reduction cost	0.63	0.063
					D_{16} Marginal emissions reduction cost	0.37	0.037
			C_7 Economic efficiency	0.1	D_{17} Total factor productivity	0.42	0.011
					D_{18} Carbon productivity	0.58	0.014
			C_8 Industrial prospect	0.4	D_{19} Industry prospect	0.22	0.022
					D_{20} Percentage of newly added capacity in industrial plan	0.63	0.063
					D_{21} Percentage of industrial GDP in regional total GDP	0.15	0.015
	Quality B_4	0.25	C_9 Trading will	0.7	D_{22} Mandatory emissions reduction	0.46	0.081
					D_{23} Marketization degree	0.28	0.049
					D_{24} Experiences in CDM	0.26	0.045
			C_{10} Relevant policies	0.3	D_{25} Substitution policy	0.32	0.002
					D_{26} Percentage of eliminated technologies/companies	0.68	0.051

Both D_{11} and D_{19} are sequencing indicators, D_{11} could be ranked in a top-down order after calculation with Formula 5.1. D_{19} could be directly traced back to the statistical yearbooks.

$$\text{Emissions reduction space} = \frac{\text{BAU - based emissions} - \text{ETS - based emissions}}{\text{BAU - based emissions}} \tag{5.1}$$

Other indicators are quantitative indicators that are available from statistical yearbooks, industry development plans, study reports, and other public channels.

(2) Model for industry selection

After the assessment indicator system is built, we shall build an industry selection model in the following three steps:

Step 1: Based on the fuzzy membership function model [6, 7], nondimensionalize the B_1–B_4 indicators. Quantify positive indicators with the upwards semi-trapezoid fuzzy membership function model (Formula 5.2). Quantify contrary indicators with the lower semi-trapezoid fuzzy membership function model (Formula 5.3).[2]

$$C(X_i) = \frac{X_i - X_{\min}}{X_{\max} - X_{\min}} = \begin{cases} 1 & X_i \geq X_{\max} \\ \frac{X_i - X_{\min}}{X_{\max} - X_{\min}} & X_{\min} \leq X_i \leq X_{\max} \\ 0 & X_i \leq X_{\min} \end{cases} \tag{5.2}$$

$$C(X_i) = \frac{X_{\max} - X_i}{X_{\max} - X_{\min}} = \begin{cases} 1 & X_i \leq X_{\min} \\ \frac{X_{\max} - X_i}{X_{\max} - X_{\min}} & X_{\min} \leq X_i \leq X_{\max} \\ 0 & X_i \geq X_{\min} \end{cases} \tag{5.3}$$

In Formulas 5.2 and 5.3, $C(X_i)$ is the membership value of the actual value, $X_{\max}$ and $X_{\min}$ are, respectively, the maximum and minimum value of D_i; the impact of a dimension could be eliminated with the fuzzy membership function model, and then subject to mathematical calculation.

Step 2: The Delphi Technique and the Analytic Hierarchy Process are used to evaluate the weight of indicators. The weight of indicator P is 1, and the weight of B_1–B_4 is the same at 1/4 (due to basically the same importance degree) based on the average weighting method.

Step 3: In terms of the industries with the weight of P above 0.65, the weight of B_1–B_4 is calculated with Formulas 5.4, 5.5, and 5.6.

[2]Positive indicator is the indicator whose higher score means better assessment result. The contrary indicator is the indicator whose lower score means better assessment result, like cost indicator.

$$T = \sum_{i=1}^{4} \sum_{j=1}^{9} \sum_{k=1}^{24} X_{ijk} W_{ijk} \quad (5.4)$$

$$Y_i = \sum_{j=1}^{9} \sum_{k=1}^{24} X_{ijk} W_{ijk} \quad (5.5)$$

$$Z_j = \sum_{k=1}^{24} X_{ijk} W_{ijk} \quad (5.6)$$

In Formulas 5.4, 5.5, and 5.6, "*T*" is the integrated score of the four modules (B_1–B_4), Y_i is the score of Module *i*, Z_j is the score of Element *j*, X_{ijk} is the weight of Base Indicator *k* (Element *j* of Module *i*) after nondimensionalization. Element C_3 and Base Indicators D_7 and D_8 are separately listed as preconditions, rather than covered in these formulas. Through the above calculations, the "*T*" value, i.e., the score marking the "industry accessibility", of each industry is finally obtained.

(3) Industry selection principles

With reference to the calculation results, we ranked the pool of industries in a top-down order based on their scores. Higher score indicates the industry is more appropriate for joining in the ETS. The global ETS experiences show that in its infancy, there should be not many regulation industries. While taking account of the fact that the Chinese overall economic development level is not high, and there is uneven economic growth among regions, it is better for the ETS to cover 2–4 industries during the trial run. In the areas that are fairly marketized and having set higher emissions reduction targets, 3–4 industries may be involved.

5.3 Methods to Define the Coverage of Guangdong Pilot ETS

5.3.1 Industry Alternatives Pool Construction

In 2010, Guangdong's gross energy consumption in all industrial sectors was 271.98 million tons (Mt) of coal equivalent; of that, 85.5% was occupied by such sectors as manufacturing, electricity, and transportation [8]. They are an epitome of all industrial sectors that use different technologies. Theoretically, all industries should be separately tested based on the above industry selection model. However, while considering data availability and cost efficiency, we streamlined the scope of sectors in this article based on the criterion that the annual energy consumption of the sector shall exceed 2% of Guangdong's total [9]; then nine sectors have won out, their combined energy consumption holds 55% of Guangdong's total (see Table 5.2).

5.3.2 To Sort out the Qualified Sectors

By referring to the metrics in Table 5.1, we assembled, collated, and calculated the data about these nine sectors, and found out that only five of them meet the preconditions, i.e., electricity, textile, petrochemistry, cement, and steel. Yet their accessibility to the ETS is subject to further assessment (see Table 5.3).

In light of Guangdong's *Plan for Adjustment and Revitalization of Petrochemical Sector*, and the layout program for key industries during the 12th Five-Year-Plan period (2011–2015), the province will quadruple the production capacity of its petrochemical sector, and place the sector at the core of industrial restructuring in the coming five years since 2011. Guangdong has great potentials to reduce carbon emissions with a large cardinal number of emissions volume. The province's petrochemical sector mostly out-purchases raw materials, but in essence, it is a highly resource-dependent sector, coupled with high percentage of occupancy for fixed assets, so it is less likely to cause carbon leakage.

Likewise, the electricity sector, with a high percentage of occupancy for fixed assets, is neither likely to cause carbon leakage. And, this sector also has great potentials to reduce carbon emissions, which is attributed to its low-carbon energy consumption mix, application of CCS and similar technologies, relatively high economic efficiency and Industry Climate Index (ICI).

In case of textile sector, it is at risk of carbon leakage if a cap was placed on its emissions volume which is too small to allow for massive reduction; moreover, it has a low ICI and small motive power to join in the ETS, which explains its lowest score among the five sectors that meet the preconditions.

In light of Guangdong's *12th Five-Year Plan for Development of Cement Sector*, its cement sector shall expedite industrial restructuring in 2011–2015. In such context, quite a number of the cement enterprises with backward capacity eliminated are able to sell the permits to emit carbon, and those holding newly operated capacity will become the buyer. In a word, the cement sector will play an active role in Guangdong ETS.

Steel sector ranks itself at a low-to-mid place in the scores of five metrics, while there are altogether nine metrics; yet the discrepancy between other sectors is still moderate, indicating that steel sector may be embraced by Guangdong ETS if its scope is extended.

Table 5.2 Sectors with separate energy consumption in 2010 holding above 2% of Guangdong's total (Unit: %)

Sector	Electricity	Textile	Paper	Petrochemical	Transportation	Cement	Construction	Steel	Livelihood	Subtotal
Percentage	7.45	3.15	3.19	8.6	10.3	4.68	2.37	4.67	11.2	55.6

Source Guangdong Bureau of Statistics. Guangdong Statistical Yearbook 2011 [M] Beijing: China Statistics Press, 2012

Table 5.3 Candidate sectors of Guangdong ETS

Sector	Preconditions C_3	Emissions C_1	Leakage C_2	Emissions reduction C_4	Technology C_5	Cost C_6	Efficiency C_7	ICI C_8	Intention C_9	Policy C_{10}	Synthetic score T
Electricity	0.984	0.101	0.042	0.028	0.039	0.053	0.035	0.064	0.014	0.035	0.411
Textile	0.689	0.015	0.021	0.052	0.013	0.044	0.024	0.017	0.012	0.027	0.225
Paper	0.604	–	–	–	–	–	–	–	–	–	–
Petrochemistry	0.989	0.124	0.051	0.089	0.045	0.062	0.027	0.069	0.052	0.015	0.534
Transportation	0.096	–	–	–	–	–	–	–	–	–	–
Cement	0.953	0.046	0.027	0.012	0.031	0.046	0.039	0.058	0.047	0.054	0.361
Construction	0.565	–	–	–	–	–	–	–	–	–	–
Livelihood	0	–	–	–	–	–	–	–	–	–	–
Steel	0.815	0.048	0.039	0.053	0.028	0.021	0.040	0.012	0.025	0.049	0.315

Note "–" indicates the concerned sector does not meet the preconditions, so they are free from further calculation

5.3.3 *Result and Conclusion*

The industry alternatives pool for the ETS will play a crucial role in Guangdong's energy saving and carbon emissions reduction in the future, yet the above analysis reveals that not all sectors with high-carbon emissions are eligible to be covered by the ETS. Only five of them meet the preconditions, i.e., petrochemistry, electricity, cement, steel and textile, which are herein placed in a descending order of their synthetic scores (see Table 5.3). Being one of China's low-carbon pilot provinces, Guangdong boasts favorable economic conditions and industry environment to implement the ETS. We hereby suggest that Guangdong ETS shall involve electricity, petrochemical, and cement sectors for the following reasons:

First, the combined energy consumption of the three sectors accounted for 20.7% of Guangdong's total in 2010, similar to the percentage within the Tokyo Protocol and qualified for the trial run of the ETS.

Second, petrochemical, electricity, and cement sectors are able to constitute a self-run carbon emissions trading market, where petrochemical sector buys the permits to emit carbon, while cement sector can sell the permits, it seems that they are able to go through the complete process of the ETS, and;

Last, Guangdong's electricity and petrochemical sectors are dominated by a small number of large-sized enterprises. In contrast, the cement enterprises are in a large number, yet they only rely on four types of cement-making kilns, which are easier for ETS administration.

Carbon emissions mainly stem from two sources: energy consumption and production process. When making a model calculation based on energy consumption, the result thereof may be somewhat deviated from the calculation based on carbon emissions volume, and the difference exists in two basic metrics as the potential for carbon emissions reduction and emissions volume, yet such deviation is of little impact on the overall calculation result.

5.4 Guangdong Total Carbon Emissions Allowances

The carbon trading competent department will, by taking account of several elements like local GHG emissions reduction targets, historical emissions of the regulation industries, and emissions reduction space of covered enterprises, set an upper limit on the emissions allowances of covered enterprises at certain point, and finally total up all upper limits on the emissions, which will be the total carbon emissions allowances. A reasonable design of total carbon emissions allowances is one of the essential elements for safeguarding ETS smooth operation.

In light of the *Notice on Carrying out the Work of Carbon Emission Trading Pilot Program in China*, and the *Program for Carrying out the Carbon Emissions Trading Pilot Work in Guangdong Province*, Guangdong ETS shall undergo a pilot period from 2013 to 2015. In order to comply with its emissions reduction obligation required by the "12th Five-Year Plan" (2011–2015), an estimation of the ETS-based total emissions allowances shall be made, then integrated with different economic growth rate estimated in scenario setting, and finally evaluate whether the ETS-regulated industries are able to fulfill the total emissions allowances of Guangdong Province.

5.4.1 Economic Growth, Energy Consumption, and Carbon Emissions

Guangdong—a fairly developed province in east China—has been a bellwether of China's reform and opening-up endeavor. With continuous economic growth, the province witnessed its 2010 GDP aggregate exceed 4.5 trillion yuan, which accounted for 1/9 of China's total, thus ranking first in the whole country. Moreover, Guangdong has kept optimizing its industrial structure, with the output value of the primary, secondary and tertiary industry holding 5.52, 51.55, and 42.93%, respectively, of its 2008 GDP, such percentage became 5.0, 50.4, and 44.6% in 2010. The ratio of the primary industry kept falling, while that of the tertiary industry was steadily going up; highlighting the predominant role of both industrial and service sectors. Guangdong attempted to control the permanent resident population at no more than 97.30 million in 2010. The province will make a continuous effort in pressing ahead with industrialization and urbanization, and sustaining economic growth. The GDP aggregate of Guangdong is estimated to reach 8.73 trillion yuan by 2020 (calculated at 2005 constant prices). At that time, the urbanization rate of registered population will reach 65%, and that of the permanent resident population will reach 74%.

In 2010, Guangdong total energy consumption reached 268 Mtce, with per unit GDP energy consumption at 0.66 tce/10,000 yuan, marking a level of energy efficiency among the front rank in China, and fulfilling the provincial overall target for energy saving and consumption reduction during the 11th Five-Year-Plan period (2006–2010). Guangdong has been insisting on developing new energy and renewables, the provincial nuclear power sector is about to release production capacity. Guangdong ranks first with a grand scale of nuclear power installed capacity: 5 GW was already completed, and 11 GW is still being built. The province is ready to vigorously develop hydropower generation projects and absorb more electricity transmitted from West China. Of the provincial primary energy consumption in 2010, raw coal, petroleum and natural gas, respectively, accounted for 48.1, 29.1 and 3.7%. Of the provincial final energy consumption, raw coal, petroleum, and natural gas, respectively, made up 11.4, 18.9, and 47%; while other types of energy held 22.7%.

Continuous industrialization and urbanization will improve the level of energy consumption of Guangdong. The industrialized characteristics of urbanization are an outcome of the rapid development of highly energy-consuming industries. The urbanization process promotes massive construction of urban infrastructure and housing, and drives up both per capita energy consumption and energy intensity. Urbanization gives rise to rigid growth in energy demand, and the emissions from increasing energy consumption have become a crucial challenge against Guangdong emissions reduction endeavor.

In light of the general arrangement of the *Program for Carrying out the Carbon Emissions Trading Pilot Work in Guangdong Province*, during the pilot period (2013–2015), Guangdong will carry out the emission allowances regulation over the companies (with annual CO_2 emissions above 20,000 tons) distributed in electricity, steel, petrochemical and cement sectors. Through verification of the companies' historical emissions and in reference to *Guangdong Statistical Yearbook 2011*, the above four industrial sectors emitted a total of about 340 Mt of CO_2 in 2010. As for the 202 ETS-regulated companies (including 62 electricity companies, 63 steel companies, 9 petrochemical companies, and 68 cement companies), their combined emissions held 80% of the total emissions of the four industrial sectors, and accounted for around 54% of Guangdong's total emissions.

5.4.2 Approach for Calculating Total Quantity of Allowances

The ETS is essentially an emission allowances trading mechanism with a cap on total emissions; thus, a reasonable estimation of total allowances is of great importance. An overly large total quantity is unable to constrain the emissions of covered enterprises, which offsets the effect of the ETS. Yet the total quantity should not be too small, otherwise, it may impede the normal production and operation of covered enterprises. The operational experiences of the ETS both home and abroad demonstrate that the total emission allowances shall be calculated with both "top-down" and "bottom-up" approaches.

Thinking of the "bottom-up" approach: verify the production and emission statistics of the covered enterprises, in order to learn about their quantity of emissions, potentials for emission reductions with the current technologies and the best accessible technologies; set different scenarios, calculate the emission reduction space of different companies distributed in all compliance industries, and add up to a total amount of allowances; evaluate the feasibility of the total amount of allowances under all scenarios (varied costs and technologies), and rehearse the allowance allocation; pay visit to industry associations and corporate representatives, discuss with them, and finally define the total quantity of allowances that are cost-effective and technically feasible.

Conduct a cross-comparison between the total emission allowances, respectively, based on "top-down" and "bottom-up" estimation approaches, take account of the emission demand and reduction potentials of the currently covered enterprises, and reserve a portion of allowances for newly operated projects and for government regulation. Through discussions among all stakeholders, a total amount of allowances will be finalized. And, the sum should be divided into two portions by their use: one is left for the covered enterprises, and the other is reserved for new projects and government regulation.

Calculation of the total amount of allowances shall abide by the following principles:

- Define the total quantity of allowances for compliance industries under the framework of fulfilling the emission reduction targets of Guangdong;
- The emission reductions of the regulated industries should not be lower than Guangdong's targets for emission reduction (a drop of 19.5%);
- Set a cap, respectively, on the emission allowances for covered enterprises and newly operated projects;
- Double counting of emissions from power generation (based on different power sources) is allowed, owing to data accessibility, and;
- Take account of the emission reduction potentials of each industry, the total quantity of allowances may not obstruct the normal operation of covered enterprises.

Overall, Guangdong ETS shall combine both "top-down" and "bottom-up" approaches for calculating the total quantity of allowances, and set a reasonable range for the total quantity.

5.4.3 Calculation Process and Result of Total Quantity of Allowances

(1) For the currently covered enterprises (top-down calculation)

Assumption: By 2015, the share of the emissions of the regulated industries in Guangdong's total emissions will remain unchanged from 2010, and Guangdong's total emissions will drop by 19.5% from 2010 levels.

Based on the above assumption and several scenarios with different GDP growth rate, we calculated the upper limit of the emissions of the currently covered enterprises by 2015. The result shows that the emission space of these companies varies greatly. With an average GDP growth rate at 8–10%, the provincial total emission allowances should be 640–660 Mtce in 2015.

Scenario A (GDP growth rate at around 8%): the total allowances of the covered enterprises will be 342 Mtce by 2015 based on their constant share in Guangdong's total emissions.
Scenario B (GDP growth rate at 10%): the total allowances of the covered enterprises will be 353 Mtce by 2015 based on their constant share in Guangdong's total emissions.

(2) For the currently covered enterprises (bottom-up calculation)

First, verify the historical carbon emissions of the covered enterprises of Guangdong ETS, process and analyze their emission data, and figure out their average emissions in past years. Second, screen the four industrial sectors involved in Guangdong ETS, and define the typical companies; investigate the production and technologies of different covered enterprises, draw up a conclusion about their BATs[3] for emissions reduction and capacity.

Third, in light of the national plan for lowering emissions by 2020, the State Council's *Work Plan for Greenhouse Gas Emissions Control during the 12th Five-Year Plan Period*, and *Guangdong Province to Carry out the National Low-carbon Province Pilot Implementation Plan*, three scenarios are set to estimate the emissions space of the covered enterprises, and add up to total allowances; with respect to cost and technical conditions, analyze the ability of these companies to handle such quantity of allowances in three scenarios. Through visits and exchanges with industry associations and corporate representatives, rehearsal and adjustment of total allowances, finalize the allowance budget that is cost-effective and technically feasible.

(3) Guangdong ETS allowance budget in 2013–2015

In light of Guangdong's plan for new installations during the 12th Five-Year Plan period, and in reference to the criteria for allocating allowances to the industries where these installations are distributed, the total emissions of these new installations by 2015 could be figured out.

Here is a formula: Guangdong ETS allowance budget = allowances to covered enterprises + allowance reserve for new installations + allowance reserve for government regulation.

The quantity of the allowance reserve for government regulation is determined by the emission reduction potentials of covered enterprises and the overall reduction targets of the ETS. Table 5.4 shows Guangdong ETS allowance budget and composition during the pilot period (2013–2015).

[3]"BAT": best available technology

Table 5.4 Guangdong ETS allowance budget in 2013–2016 (Unit: Mt)

Timeframe	Trading period	Allowance budget	Compliance companies	Allowance reserve	
				New installations	Government
Pilot period	2013	388	350	20	18
	2014	408	370	38	
	2015	408	370	38	
Transitional period	2016	386	365	21	

Source Guangdong Provincial Development and Reform Commission, *Notice of the First Allowance Allocation Plan in Guangdong ETS*. 2013.11.25 [10]
1. The *Notice on Carrying out the Program for Guangdong Emission Allowances Allocation in 2016* (No. 430 [2016] GD DRC Climate Office) [EB/OL]. GD DRC official website: http://210.76.72.13:9000/pub/gdsfgw2014/zwgk/tzgg/zxtz/201607/P020160708579091026333.pdf
2. The *Notice on Carrying out the Program for Guangdong Emission Allowances Allocation in 2015* [EB/OL]. GD DRC official website: http://210.76.72.13:9000/pub/gdsfgw2014/zwgk/tzgg/zxtz/201507/P020150713592839853460.pdf
3. The *Notice on Carrying out the Program for Guangdong Emission Allowances Allocation in 2014* (No. 495 [2014] GD DRC Climate Office) [EB/OL]. GD DRC official website: http://210.76.72.13:9000/pub/gdsfgw2014/zwgk/zcfg/gfxwj/201503/t20150309_305043.html
4. The Notice on Carrying out the Work for Guangdong First Emission Allowance Allocation (For Trial) (No. 3537 [2013] GD DRC RES & ENV) [EB/OL]. GD DRC official website: http://zwgk.gd.gov.cn/006939756/201401/t20140123_463411.html
5. Guangdong Provincial Development and Reform Commission, *Notice of the First Allowance Allocation Plan in Guangdong ETS*. 2013.11.25. http://www.gddpc.gov.cn/xxgk/tztg/201311/t20131126_230325.htm

References

1. Contribution of Working Groups I, II and III to the Fourth Assessment Report of the Intergovernmental Panel on Climate Change. Core Writing Team, Pachauri, R.K. and Reisinger, A. (Eds.) IPCC, Geneva, Switzerland. pp 104.
2. Yigang Wang, Xing'an Ge, et al., China's Pathway towards Carbon Emissions Trading Scheme [M]. Beijing: China Economic Publishing House, 2011: 113.
3. Jian Liu. Introduction of Germany Carbon Emission Trading Program [N]. China Environment News, 2006-08-11.
4. Wenjun Wang, Daiqing Zhao, Chonghui Fu, Analysis on the International Regional GHGs Emission Trade System's Applicability to the Province-level ETS in China and Policy Suggestion, Bulletin of Chinese Academy of Sciences, 2012, 27 (5): 602–610.
5. Wenjun Wang, Daiqing Zhao, Chonghui Fu, How to Define the Scope of China Regional CO_2 ETS Coverage: Experience and Case Study: Taking Guangdong Province as Example [J]. Ecological Economy, 2012, 7: 70–74, 82.
6. Shuhua Lu. Social Statistics [M]. Beijing: Peking University Press, 2010.
7. Lulu Zhu, Lazhen Xiao, Establishment of Comprehensive Assessment Index System for "Two-oriented Society" and Empirical Analysis [J]. Journal of the Postgraduate of Zhongnan University of Economics and Law, 2009, 6: 83–90.
8. The Economic & Information Commission of Guangdong Province, *The Notice on Guangdong Province's 12th Five-Year Plan for Energy Saving* [EB/OL].2011 [2011-10-07] http://www.gdei.gov.cn/flxx/jnjh/zcfg/201107/t20110721_105234.html.

9. Guangdong Bureau of Statistics. *Guangdong Statistical Yearbook 2011* [M]. Beijing: China Statistics Press, 2011.
10. Guangdong Provincial Development and Reform Commission, *Notice of the First Allowance Allocation Plan in Guangdong ETS*. 2013.11.25. http://www.gddpc.gov.cn/xxgk/tztg/201311/t20131126_230325.htm.
11. The *Notice on Carrying out the Program for Guangdong Emission Allowances Allocation in 2016* (No. 430 [2016] GD DRC Climate Office) [EB/OL]. GD DRC official website: http://210.76.72.13:9000/pub/gdsfgw2014/zwgk/tzgg/zxtz/201607/P020160708579091026333.pdf.
12. The *Notice on Carrying out the Program for Guangdong Emission Allowances Allocation in 2015* [EB/OL]. GD DRC official website: http://210.76.72.13:9000/pub/gdsfgw2014/zwgk/tzgg/zxtz/201507/P020150713592839853460.pdf.
13. The *Notice on Carrying out the Program for Guangdong Emission Allowances Allocation in 2014* (No. 495 [2014] GD DRC Climate Office) [EB/OL]. GD DRC official website: http://210.76.72.13:9000/pub/gdsfgw2014/zwgk/zcfg/gfxwj/201503/t20150309_305043.html.
14. The Notice on Carrying out the Work for Guangdong First Emission Allowance Allocation (For Trial) (No. 3537 [2013] GD DRC RES & ENV) [EB/OL]. GD DRC official website: http://zwgk.gd.gov.cn/006939756/201401/t20140123_463411.html.

Chapter 6
Guangdong Pilot ETS Allowances Allocation Mechanism

6.1 Foreign and Chinese Emission Allowance Allocation Criteria

The "cap-and-trade"-based ETS follows two principles for free allowance allocation: "Grandfathering" and "Benchmarking", which apply to different conditions and objects [1]. Grandfathering is a free allocation based on average historical emissions of covered enterprises. It is applicable to the companies (e.g., in chemical and electronics sectors) that have a complex process, multiple categories of technologies, and a great variety of products. And, it costs much to gather their emission data. In the case of Benchmarking Allocation, the allowances are the production/value of a working procedure or a product, which is multiplied by unit emissions.

The above basic approaches give rise to several mixed allocation approaches, e.g., "Historical Benchmarking" and "Correction Benchmarking"; the former calculates the companies' allowances based on their historical production/value and unit emissions; the latter differs from the former in a procedure for compliance period accounting, i.e., when the allowances for each compliance period are about to be settled, the competent departments will reverify the year's allowances by taking account of the companies' actual production/value, and following the principle of "any excess payment shall be refunded and deficiency shall be repaid."

During the prior two trading phases of EU ETS, most Member States adopted the Grandfathering Allocation, which was replaced by the Historical Benchmarking during the third trading phase after the carbon trading system was improved and more data were accumulated. Such benchmarks are made up of product benchmark, heat benchmark, fuel mix benchmark, and process emissions benchmark (listed in order of priority). The product benchmark involves the benchmark value for 52 products in 21 sectors. The CAL ETS also applies the Historical Benchmarking, which incorporates product benchmark and energy benchmark. The product benchmark involves 28 products in 18 subsectors. With respect to China's seven

© China Environment Publishing Group Co., Ltd. and Springer Nature Singapore Pte Ltd. 2019
D. Zhao et al., *A Brief Overview of China's ETS Pilots*,
https://doi.org/10.1007/978-981-13-1888-7_6

carbon markets, they adopt Grandfathering, Historical Benchmarking, and Correction Benchmarking in light of their respective characteristics.

Grandfathering is a common allocation approach and fairly practicable, yet it seems unfair for the companies with low emissions or already enforced emission reduction measures, thus giving rise to the so-called "whipping the fast and hard-working". Moreover, once the base year for the Grandfathering Allocation is defined, it should not be altered; otherwise, the covered enterprises may lose motivation for cutting emissions, leaving the carbon market to exist in name only. In order to incentivize the emitters to reduce emissions, the depreciation factor shall be introduced in reference to base year's emissions, and gradually lower the emission allowances to the companies.

The definition of a base year is fairly important for both Grandfathering and Historical Benchmarking, because different base years may diversify allowance quantity and emission reduction effect, which calls for comprehensive data about the companies' historical emissions and operation. Since these two approaches adopt static historical data, whenever there are economic fluctuations, the covered enterprises' production and emissions may deviate greatly from their historical records, which will then cause serious surplus or deficiency of emission allowances, drastic carbon price swings, and finally impede smooth operation of the carbon market.

The Correction Benchmarking is able to avert such problem, and the companies will exempt from meeting the total allowances through reducing production based on Grandfathering. The benchmarks could be emissions from unit production and emissions from unit value.

Emissions from unit production equal to unit consumption or benchmark value, the value may be industrial benchmarks or average value, global, national, or local, the average value of the covered enterprises (or the companies in the top rank), or historical unit consumption of each company.

In light of the working procedures and diversity of products of different sectors, set a benchmark value of a working procedure (e.g., the clinker working procedure of cement sector and long-flow process in iron and steel sector), or set a benchmark value for a product (e.g., the benchmark value for the 52 products during the third trading period of EU ETS).

If there are overly many differences between companies, or there are no data accumulation about carbon emissions for unit production, emissions for unit value (e.g., the manufacturing companies involved in Shenzhen ETS). Such allocation approach presents higher requirements for carbon verification. As such, Shenzhen Carbon Emission Exchange established accounting audit for carbon verification.

6.2 Foreign and Chinese Allowance Allocation Approaches

(1) Free allocation

Free allocation is at present the most broadly applied by the world carbon emissions trading schemes at the initial phase, i.e., the government will, after setting a cap on the national or local total emissions, compute the emissions space of the covered enterprises in line with certain principles and criteria, and then grant the emissions permit to the covered enterprises for free (usually called emission allowances) to sustain their normal operation and production. It is, in fact, an instrument for directing the participants to realize low-carbon production process.

The rationale behind the free allocation is as follows: if the free allowances fail to cover the companies' actual emissions, they have to pay for additional emission permit; if the companies cut their emissions below the allowances through active reductions, they are entitled to sell the surplus allowances to gain profit. It is a common practice for the world carbon emissions trading schemes to choose free allocation at the infant phase. Take EU ETS—the world largest "cap-and-trade" system—for instance, most of the allowances were allocated for free during its prior two trading periods (2005–2007 and 2008–2012); the percentage of free allowances will go down to approximately 50% during the third period (2013–2020). Free allocation also prevails in other countries and regions, e.g., CAL ETS allocates free allowances to the industrial sector that is exposed to significant risk of carbon leakage, NZ ETS adopts the same attitude toward the forestry, fishery, and industry that are prone to be affected by international competition, as well as China's seven pilot carbon markets.[1]

(2) Fixed-price sales

Allowance sales are a supplement to free allocation. The government usually sets a price for emission allowances. In case the free allowances are not enough for the covered enterprises to sustain normal production, they are able to purchase additional allowances from the government at a fixed price.

During the initial operation of EU ETS, the carbon prices experienced marked ups and downs, which is a lesson for its counterparts in other countries. For instance, AU ETS adopts a strategy of "progressive marketization", i.e., sell the allowances at a fixed price in the first 3 years, allow price fluctuation within a permissible range in the coming 3 years, and ultimately apply market pricing, for the purpose of sending a stable price signal to meet the anticipation of all participants, give them enough time for adjustment and adaptation, or even divert funds and capital to flow into emission reduction projects. Similarly, NZ ETS specially

[1]With the exception of Guangdong ETS that also adopts allowance auction, the other six pilot carbon markets mainly rely on free allocation.

arranges a transitional period, where the government sells the additional allowances at a fixed price of 25 NZ$/t.

(3) Auction

Paid allocation of allowances is mainly realized through auction. There are a sealed auction and dynamic auction based on rounds of biddings. Currently, it is sealed auction that prevails in the current carbon markets. The dynamic auction is applicable to mature traded products, it is unable to incentivize the entrants to the newborn carbon market, because the potential bidders who are less competitive may be deterred from participating; moreover, it is possible to breed conspiracies. Such aftermath will distort the healthy price signal and allocation results.

The sealed auction is made up of uniform-price auction and discriminatory auction. In case of uniform-price auction, bid winners pay for the allowances at the uniform market-clearing price. In case of a discriminatory auction, bid winners pay for the allowances at their own bidding prices, which are diversified and above the uniform market-clearing price [2]. Uniform-price auction is easier to operate, and able to draw more bidders and promote reasonable competition. In contrast, in case of a discriminatory auction, the smaller and unsophisticated bidders are prone to think that it is hard to predict the market-clearing price, so they dare not join in the bidding for fear of false judgment. Particularly at the initial phase of the ETS, the secondary market still lacks mobility, and there are uncertainties in sending a stable price signal. Summing up, uniform-price auction is more appropriate for increasing participating companies; specifically, uniform-price sealed auction is deemed as the best option for initial allowance allocation.

EU ETS distributes more and more allowances via auction during the first three trading periods. The Member States obtain at most 5% of allowances via auction in the first period, such proportion goes up to 10% in the second period, and some sectors may receive allowances solely via auction in the third period. During the three periods, the auction-based allowances granted to different sectors are in various proportions. For instance, EU ETS distributes 100% of the allowances to electricity sector via auction by 2020, such proportion for other sectors goes up year by year to reach 70% by 2020, and ultimately reaches 100% by 2027. The US Regional Greenhouse Gas Initiative (RGGI), which has been implementing allowance auction since it was launched in 2003, is the world's first cap-and-trade system that solely relies on allowance auction.

In contrast to free allocation and fixed-price sales, allowance auction boasts lots of advantages, e.g., simplified procedure, explicit rules, clear-cut targets, fair, and cost-effective, which are fit for exploring authentic prices for emission credits. Allowance auction is able to motivate the companies to cut emissions more actively, create a fairly competitive environment for the new entrants, promote the application of the latest production technologies, and ultimately increase energy-use efficiency and lower emissions. Furthermore, the government may use the gains from allowance auction to sponsor the emission reductions in underdeveloped areas, foster R&D, and application of clean technologies, or subsidize the carbon-intensive export-oriented companies to tackle with international competition.

The biggest resistance in front of emission credit auction is actually from the ETS-regulated companies, since all of them will suffer mounting emission cost. In order to urge the companies to cut emissions conscientiously, there shall be explicit laws and regulations, the companies shall form strong low-carbon awareness and properly use the income from emission credit auction. Besides, the auction mechanism has intrinsic problems that cannot be ignored. Since different sectors and companies have greatly varied total emissions and reduction potentials, a cover-all auction mechanisms seems unfair for some of them, i.e., the privileged sectors or large companies may monopolize credit trading via auction, thus depriving the vulnerable sectors or small companies of their advantages.

(4) Combined allocations

The above allocation approaches have their respective strengths and shortcomings and the best applicable industries. If an emissions trading scheme involves a large number of sectors and greatly distinguished emitting installations, it usually combines the two allocation approaches to make use of their advantages. Free allocation is usually preferred by a newborn carbon market, allowance auction is then gradually introduced after the market matures. Alternatively, an ETS is able to combine both free allocation and auction at its infant phase.

In its first and second trading periods and early stage of the third period, EU ETS combines free allocation and allowance auction, but gradually increases the auctioned allowances.

CAL ETS learned from the experiences of EU ETS, and then refined the methodologies for allowance-free allocation and auction. The competent department is entitled to auction 10% of the allowances directly, leaving the rest 90% allocated to companies for free; yet it asks the companies of different categories to handle the allowances differently. For instance, the industrial sectors and independent power plants can use the free allowances directly; the Publicly Owned Utilities (POU) may auction their surplus emission credits voluntarily; the Investor Owned Utilities (IOU) shall auction all of their free allowances, and return the income therefrom to electricity consumers.

The ETS operational experiences in Europe and America prove that the partially auctioned allowances are more effective in boosting the companies to cut emissions and energize the carbon market.

6.3 Foreign and Chinese Allowance Allocation Frequency and Effectiveness

For the purpose of making the ETS administration more compatible with the actual situations of an economic society, the allowances are allocated in prescribed time frames, and different allocations are chosen in different time frame, e.g., allocation by year or by stage, different allocation approach decides the validity period of allowances, which also gives rise to the issues of allowance storage and loaning.

The European Commission provides that the EUAs can be either stored or loaned within the same trading period. The EUAs are distributed annually, but every date of issuance is 2 months ahead of the deadline of the previous compliance year, which implies that if a company is short of EUAs in 1 year, it may use next year's allowances. By analogy, if a company has surplus EUAs, they may store them for next year within the same trading period, rather than saving them for the next period. As for China's seven pilot carbon markets, their allowance allocation systems vary from each other. For allocation frequency, Shanghai ETS allocates 3 years' allowances all at once, but the covered enterprises shall fulfill their annual obligations on time; other ETSs allocate the allowances annually and require fulfillment of annual obligations; the allowances in all ETSs are available for storage, rather than loaning; the allocation methodologies are basically uniform for all sectors (these seven markets abide by the Benchmarking Principle for allocating allowances to electricity sector, but set up different allocation criteria).

Allowance banking, on one hand, incentivizes the companies with low reduction cost to further reduction efforts and start earlier, and on the other hand buffers the short-term impact from varying demand and mitigate allowance price fluctuations. But the banking is valid for the same period, cross-period banking is not allowed unless the carbon market becomes fairly mature and stable. The operational experiences of the existing ETSs reveal that the target for capping total emissions seems unreasonable, because of lacking historical data and experiences in projecting emissions. At this point, the main objective of ETS is to obtain accurate emission data and operational experiences, a ban on cross-period allowance banking is a better option. After the ETS becomes more mature, the upper limits on total emissions will be coherent, and cross-period allowance banking may receive a better result.

Allowance borrowing may lure the companies to shift the emission reduction pressure to the last year of a trading period, which is not a good news for either administration of the covered enterprises, or for discovering the problems existing in the carbon market at an early date. Currently, on China's pilot carbon markets, it is government reserved allowances that take charge for easing the situation of short allowances.

6.4 Guangdong ETS Allowance Allocation

6.4.1 Interpretation of Guangdong ETS Allowance Allocation Plan

Guangdong Province Development and Reform Commission (GD DRC) takes charge of the overall work of the Guangdong Carbon Emissions Trading Scheme, e.g., allowance administration, monitoring, and guidance. In order to work out a scientific and appropriate allowance allocation plan, GD DRC, in conjunction with the relevant departments, established the Allowance Allocation Assessment Panel and Industrial Allowance Assessment Review Board to evaluate the ETS allocation methodologies, criteria, and rationality, the concrete work may be undertaken by the accredited social organizations or institutions. The development and reform commissions at or above the prefecture level are obliged to coordinate with GD DRC in allowance administration. They are responsible for reviewing the allowance purchase application filed by the new project proponents under their jurisdiction, and urge the covered enterprises under their jurisdiction to fulfill their obligations.

Fairness and efficiency are core elements of an allocation mechanism, but their status may not be equal under different policy goals, which leads to greatly varied allocation plans. Before an allocation plan is determined, the policy goal and guideline for each trading period shall be clarified. A pilot carbon market is obliged to make explorations for the ETS, and compatible with the local emission reduction targets and industrial structure over 2011–2015 (the Five-Year Plan period). If an ETS covers a great variety of sectors, the corresponding allocation plan shall stress allocation efficiency, like what Shenzhen ETS has done; otherwise, a fairer allocation plan is preferred. Guangdong ETS has been following the guideline as "efficiency comes first, balances fairness, promotes emission reduction and safeguards development," which is interpreted as follows:

(1) Promote emission reduction and safeguards development: Stick to the overall target for controlling total emissions, and in the meantime leaves appropriate space for continuing steady development of socioeconomy. Guangdong is China's top developed province; its economic aggregate and growth have been ranking first since China introduced the reform and opening-up policy in 1978. However, it should be noted that the entire Guangdong is unevenly developed, the gap between the developed areas and underdeveloped areas is even wider than such a gap across the nation, i.e., Guangzhou and Shenzhen within Guangdong's jurisdiction are the top developed cities in China, yet the counties at both the eastern and western wings and in mountainous areas are the nationally designated poor counties. As such, Guangdong ETS shall make effort to fulfill the target for controlling total emissions, and pay attention to safeguard the development of the less developed areas.

(2) Control existing emissions and restrict emission increments: Motivate the existing covered enterprises to cut emissions step by step, and set strict emission criteria for new project proponents. The existing production facilities can only make limited emission reductions, because massive reduction calls for energy-saving transformation, which takes a long time to finish and incurs higher expenses, it is impossible for most of the existing covered enterprises to upgrade and update all equipment within the pilot period; instead, they rely on administration of energy conservation and consumption of low-carbon materials to cut emissions. Overall, a reasonable allocation plan shall ask the existing covered enterprises to gradually lower their emissions so as to control their current total emissions; while the new projects do not have such technology or equipment lock-in, they shall satisfy stricter allocation criteria to become advanced emitters.

(3) Efficiency comes first, balances fairness: Take account of both the benchmark emissions by sector and historical emissions by company, support the advanced sectors/companies, and winnow out the outdated ones. Emission allowance allocation is actually a configuration of development space. A proper allocation plan shall take full consideration of the local industrial development policies, take a strict position on the companies "blacklisted" by the government due to high emissions, and high pollution, but incentive those sunrise industries. Different sectors have greatly distinguished emissions, some of them, with fairly low emissions, could be compared to the international advanced level; while some of them have middle-level emissions. Therefore, when developing the allocation criteria by sector, we should take an overall account of the sector's unit CO_2 emissions, and then come up with a set of allocation criteria that satisfy the emission demand of most companies.

(4) Paid allocation and implementation step by step. Rely on free allocation with paid allocation as a supplement, and gradually increase the proportion of paid allocation. According to the *Report to the Eighteenth National Congress of the Communist Party of China* delivered by the then China's president Hu Jintao on 8 November 2012, China shall deepen the resource product price and tax-fee reform, establish the paid use system of resources and eco-compensation system that are able to reflect resource supply–demand pattern, resource scarcity, ecological value and intergenerational compensation. Paid allocation of emission allowances is an irresistible trend. Both EU ETS and RGGI have started paid allocation in an all-round manner, and Guangdong ETS has been making attempts in this regard. However, for the sake of lowering emission cost of companies and reducing ETS operational resistance, Guangdong ETS still relies on free allocation, only a small portion of the allowances are distributed via auction with an aim to send a signal that resource has its price.

6.4.2 Allowance Allocation Methodology

Guangdong Development and Reform Commission (GD DRC) takes charge of drawing up emission allowance allocation plan for Guangdong ETS, by taking full account of the baseline emissions, reduction potentials, and historical emissions by sector. Such plan shall be first reviewed by Guangdong ETS Allowance Assessment Review Board, and then reported to and ratified by Guangdong People's Government. GD DRC had issued three allowance allocation plans respectively for 2013, 2014, and 2015, which define the covered enterprises, new projects, annual total allowances, and percentage of free and paid allowances, allocation principles, methodologies, and procedures, allowance quantity via auction, transaction platform, and rules.

Guangdong ETS allowances are made up of the compliance allowances and reserve allowances; the latter is left for new entrants (to satisfy their emissions demand) and for the government (to regulate carbon market in case of allowances shortage). Guangdong ETS allowances are subject to both free and paid allocation. The allowances for new entrants are estimated in light of their production capacity and the allocation criteria. The allowances for market regulation are determined in certain proportion to the compliance allowances.

(1) Allocation plan

An allowance allocation plan involves main body, schedule, channel, procedure, and price. Generally, the main body for allowance allocation refers to the competent government department. Take Guangdong ETS, for instance, it is under the leadership of Guangdong Development and Reform Commission (GD DRC); its allowances allocation is in line with the Guangdong ETS allowance allocation plans, respectively, for the compliance year as 2013, 2014, and 2015.

Guangdong ETS adopts both free and paid allocation. In order to coordinate the work for free and paid allocation, they are scheduled at different times and through different channels. In principle, GD DRC shall distribute the free allowances to Guangdong ETS Allowance Registration System on every July 1, the covered enterprises shall register an account in this system and apply for 95–97% of these free allowances. Paid allocation is undertaken by an auction platform titled China (Guangzhou) Emissions Exchange in each quarter, both covered enterprises and new entrants can bid for the allowances via this platform at their own emissions demand. Guangdong ETS allowance allocation plan is briefed in Table 6.1.

Owing to a limited pilot period (2013–2020), Guangdong ETS allowances are allocated annually for two aspects of considerations: on one hand, such allocation shall enable the administrator to acquaint with the administration process and gain a thorough understanding of the allocation mechanism; on the other, it shall favor adjustment to the allocation plan, and leave a time window for such adjustment after a period of market observation and feedback.

If the basic data are fairly complete, the allowance allocation shall center on emitting installations to prevent from wrong identification of regulatory boundary.

Table 6.1 Essential elements of Guangdong ETS allowance allocation plan [3–5]

		Free allocation (compliance companies)	Free allocation (compliance companies, new entrants)
Schedule		Each July 1	Each quarter
Channel		Allowance registration system	Guangzhou Emissions Exchange
Approach		Direct distribution	Bidding
Price		Free	Floor price
Quantity	2013	97% of total verified allowances	Purchase 3% of total verified allowances
	2014–2015	95% of total verified allowances for electricity sector; 97% for other sectors	Purchase at their demand

Source The Guangdong ETS allowance allocation plans, respectively, for the compliance year as 2013, 2014, and 2015

At the initial phase of Guangdong ETS, the basic data about covered enterprises are incomplete, since lots of them do not store any emission data about emitting installations, thus making companies become the focus of allowance allocation.

In order to define the regulatory scope or boundary, the covered enterprises shall submit an emissions monitoring plan to the administrator for examination and approval, report their emissions data based on the plan. Later, based on the report and allocation criteria, the administrator shall calculate the allowances to be distributed.

(2) Allocation criteria

After a thorough comparison of the Benchmarking and Grandfathering Principles, Guangdong ETS gives priority to the Benchmarking Principle, and turns to the Grandfathering Principle when the Benchmarking is inappropriate. For instance, the Benchmarking Principle is applicable to the single generating set in the electricity sector, the clinker production and grinding in cement sector, and the long process in iron and steel sector. In contrast, the Grandfathering Principle is applicable to co-generation in electricity sector, mining of cement mine, oil refinery, and ethylene production in petrochemical sector, a short process in iron and steel sector, and other sectors with diverse types of products and sophisticated process.

The allowance allocation criteria of Guangdong ETS are briefed as follows. The years' allowance allocation plans are available for downloading from the website of Guangdong Development and Reform Commission (http://www.gddrc.gov.cn/was5/web/search) for the readers that care about Guangdong ETS.

(i) Electricity sector

In light of Guangdong ETS allowance allocation plan, the emission allowances are granted to electricity sector by following the Benchmarking Principle.

The electricity sector that falls under regulation by Guangdong ETS consists of electricity generation companies and electricity–heat co-generation companies. The generation companies of the same category have different types of generating sets which are made up of those based on coal and natural gas. The coal-fired generating sets are further divided into four subcategories based on the size of installed capacity; the big and small generating sets have greatly varied energy consumption and emissions owing to varied technical parameters. Therefore, the allocation criteria are established on basis of the average energy consumption of the same-category generating sets, which not only reveals the emissions disparity among various generating, but motivates the companies with the same fuels and generating sets to seek energy-efficiency modification or automatic elimination.

a. Electricity generating units

The total allowances granted to covered enterprises are made up of free and paid allowances; the former is computed in light of allocation criteria, while the latter is decided by the companies at their special demand. Therefore, the percentage of paid allowances shall be ignored when setting up the allocation criteria. Formula 6.1 is the allocation criteria for electricity generation companies.

$$\mathrm{EA} = \sum_{i=1}^{n} \mathrm{HP}_i \times \mathrm{BM}_i \times \mathrm{CF} \times (1 - \mathrm{PA}) \tag{6.1}$$

EA is free allowances allocated to the companies with generating set i; HP_i is the electricity output of generating set i; BM_i is benchmark emissions from generating set i of the same category; CF is annual decline factor; PA is the percentage of paid allowances.

HP_i is electricity output of generating set i (average historical output in 2013 allocation, and actual output in 2014 and 2015 allocation). It is provided that the electricity output for allowance computation may not be larger than the designed capacity of the generating set. BM_i is determined on basis of the unit emissions of all types of generating sets. The allocation criteria are finally determined in reference to the emissions reduction target (total allowances) of the electricity sector. See the benchmark emissions of Guangdong ETS-regulated electricity generation companies in Table 6.2.

b. Heat-electricity generating units and resources comprehensive utilization generating units[2]

The Grandfathering Principle is applicable to the allowance allocation for the heat–electricity generating units and the resources comprehensive utilization generating unit, because the former has greatly varied heat–electricity ratio and inadequate data accumulation; the latter has low energy use efficiency, fuel grade notably lower than the generating units based on general fossil fuels.

[2]Resources comprehensive utilization generating units are those mainly rely on coal gangue and oil shale as fuel.

Table 6.2 Benchmark emissions of Guangdong ETS-regulated electricity-generating sets

Category of generating sets			Benchmark emissions (gCO_2/kWh)	
			2013	2014 and 2015
Coal-based	1000 MW		770	825
	600 MW	Ultra Super-critical (USC)	815	850
		Super-critical (SC)		865
		Sub-critical (SubC)		880
	300 MW	Non-circulating fluidized bed (NCFB)	865	905
		Circulating fluidized bed (CFB)		927
	<300 MW	Non-circulating fluidized bed (NCFB)	930	965
		Circulating fluidized bed (CFB)		988
Natural gas-based	390 MW		415	390
	<390 MW		482	440

Source The Guangdong ETS allowance allocation plans, respectively, for the compliance year as 2013, 2014, and 2015

$$\text{Companies}' \text{ allowances} = \text{historical average emissions} \times \text{CF} \times (1 - \text{PA})$$

(ii) Cement sector

The production of cement follows three fundamental stages as quarrying, clinker production, cement grinding, and micro-grinding. The total allowances granted to cement plants are a sum of allowances for the four stages. The Benchmarking Methodology applies to clinker production and cement grinding, while the Grandfathering Methodology applies to quarrying and micro-grinding.

Similar to the calculation formula for the allowances for electricity sector, HP_i is cement production (average historical output in 2013 allocation, and actual output in 2014 and 2015 allocation). Since cement overproduction happens from time to time, it is provided that the cement production for allowance computation may not be 1.3 times more than the designed capacity of the cement kiln. The categorization and value of BM_i are given in Table 6.3.

Table 6.3 Benchmark emissions of Guangdong ETS-regulated cement sector

Benchmark emissions (CO_2/t)	Cement clinker production line			Cement grinding
	$\geq$ 4000 t/d	2000 $\leq$ CO_2/t < 4000 t/d	<2000 t/d	
2013	0.874	0.921	0.945	0.0237
2014 and 2015	0.893	0.937	0.950	0.027

Source The Guangdong ETS allowance allocation plans, respectively, for the compliance year as 2013, 2014, and 2015

Clinker production efficiency is in direct proportion to the capacity of the production line. In order to mobilize the initiative of cement producers, Guangdong ETS exercises differentiated allowance allocation criteria toward different clinker production lines. For the production lines at or above 4000 t/d, the BM_i is the unit average emissions from the producers of the same technical group. For the production lines no less than 2000 t/d and less than 4000 t/d, the BM_i is the unit average emissions from the leading 75 producers of the same technical group. For the production lines less than 2000 t/d, the BM_i is the unit average emissions from the leading 50 producers of the same technical group.

(iii) Iron and steel sector

Iron and steel sector is made up of long-flow and short-flow iron and steel companies. In this study, the focus will be placed on long-flow companies. Their coke and lime input (raw materials) and crude steel production (product) are directly related to CO_2 emissions. The Benchmarking Principle is herein feasible.

Similar to the calculation formula to the allowances for electricity sector, HP_i is raw material input or crude steel production i (average historical input/production in 2013 allocation, and actual input/production in 2014 and 2015 allocation). In order to safeguard the fairness of allowance allocation, it is provided that the steel production for allowance computation may not be 1.1 times more than the designed capacity. The categorization and value of BM_i are given in Table 6.4.

The BM_i of the existing long-flow iron and steel companies in Guangdong is defined through weighted calculation. The carbon-bearing substances directly emitted from crude steel production is made up of scraps, slags, steel, pig iron, lime, dolomite, carburant, and coke. For lime roasting, the emissions are from decarbonation and energy consumption. The emissions from coking process are beyond consideration because the exhaust from coking is inflammable gas with high heat value and recyclable, the coking by-products as benzene (solid) and coal tar (liquid) are either solid or liquid, and the coke will be directly filled in the iron-making furnaces.

Summing up, the emissions hereby are from energy consumption during coal preparation, coking, and gas purification, after deducting the energy for self-recycling and for supply to external costumers.

Table 6.4 Benchmark emissions of Guangdong ETS-regulated long process in iron and steel companies

Raw material/Product	Benchmark emissions (CO_2/t)	
	2013	2014 and 2015
Coke (raw material)	0.3307	0.2976
Limestone (raw material)	0.8280	0.5796
Crude steel (product)	2.1274	1.9785

Source The Guangdong ETS allowance allocation plans, respectively, for the compliance year as 2013, 2014, and 2015

(3) Allocation approaches and procedures

There are free and paid allowance allocations. Paid allocation a unique feature of Guangdong ETS, its approaches and procedures will set an example for the ETS in other areas and the nationwide uniform carbon market.

(i) Receivers of allowances and allocation approaches

As soon as the pilot work for Guangdong ETS started in December 2013, the paid allocation was put into effect, catering to both covered enterprises and new entrants. As approaching the end of the first compliance year 2013, a paid allocation system was established and started the smooth trial run. In addition to covered enterprises and new entrants, there were other opt-in entities. Therefore, since the second compliance year 2014, the investment institutions and other market participants are able to join in an auction for paid allowances.

For covered enterprises, the legal persons are entitled to determine the quantity of free allowances, and whether and when to auction for paid allowances based on their emissions demand.

For new entrants, the owners shall, before the projects are completed and subject to acceptance check, purchase adequate verified paid allowances at China (Guangzhou) Emission Exchange (CGEX). After the projects are put into operation for 12 months, the new entrants will access the same allocation methodologies as the covered enterprises.

(ii) Allocation procedures and rules

a. Free allocation

The competent department of Guangdong ETS will, in reference to the basic emissions data supplied by the covered enterprises, calculate their respective free allowances based on allocation criteria, submit the result to the Allowance Allocation Review Board for rectification, and then issue the allowances to the companies' registered account at CGEX.

After these companies have fulfilled current year's obligations, the competent department will issue next year's allowances to their registered account. The allowances used for covering current year's emissions will be revoked, and the surplus allowances are still valid. The benchmark values of these companies will directly affect the quantity of allowances, therefore, when defining the benchmark values, the competent department shall take account of production stability and sustainability of these companies, so the benchmark values for the year 2013 are an average of the verified emissions data over 2010–2012.

The allowances given to the companies with abnormal production are defined in line with the two principles as follows: First, the projects, either newly operated or going through substantial technical revamp, fail to reach the prescribed period of operation. If such projects are operated within 1 or 2 years, or if the technical revamp brings significant changes upon their original categorization, production process, or capacity, their allowances shall be based on the emissions from the period of normal production, for the sake of fairness. The specific rules are given like this, if the projects fail to reach 12-month operation in a current year, then the allowances shall be based on the average emissions of a full year, or in reference to the average historical emissions from the counterpart companies. Second, special circumstances may take place during the production process within the years for drawing the basic data. Any inappropriate production management, external natural environment, and social environment may lead to abnormal production, which will then bring uncertainties upon allowance determination. When deciding the quantity of allowances, full consideration shall be given to the reasons, duration, and impact of abnormal production. If the abnormal production arises from changes on demand–supply pattern or equity structure, or from technical revamp, all the risks therefrom are due to borne by the companies themselves, so allowance determination shall follow the regular procedure, in other words, there is no special treatment.

Special consideration is given to the following two situations: Any abnormal production arises from force majeure, e.g., typhoon, earthquake, flood, fire, or explosion, and extends over 1 month. Any abnormal production under government mandate (excluding electricity generation dispatching) and extends over 1 month. Under such circumstances, the allowances may be calculated disregard of current year's emissions, or converted based on the emissions from normal production.

a. Paid allocation

From December 2013 to June 2014, China (Guangzhou) Emission Exchange altogether held five rounds of allowance auctions with reserve price determined by the government. Such an auction reserve price in 2013 was set at 60 yuan/tCO_2. This is a sealed auction at uniform price. The companies shall declare their intended purchase quantity and price via the electronic transaction system. Upon closure of the auction, the transaction system shall rank the declarations by price, then rank the same-priced declarations by time. If the total declared quantity is no more than the total issuance quantity, all the transactions will be concluded at declared quantity at the reserve price. If the total declared quantity is above the total issuance quantity, then all the transactions will be concluded at issuance quantity at the minimum offer price.

Starting from the second compliance period of Guangdong ETS, the competent department decided to hold an allowance auction at the last month of each quarter. In the compliance year of 2014 (from September 2014 to June 2015), the auctions were exercising a staircase reserve; such price for the four auctions was, respectively, 25, 30, 35, and 40 yuan/tCO_2. In the compliance year of 2015 (from September 2014 to June 2016), the auction reserve price was 80% of the weighted average transaction price for the listed tradings at CGEX during the 3 calendar months before the Notice Date when the auction price is made to public. The companies shall declare their intended purchase quantity and price via the electronic transaction system. They are allowed to make multiple offers with each one no less than 1000 tCO_2; no offer may be canceled after submission. And, the accumulated offers may not be over 50% of total issuance quantity. The auction is a uniform-price auction, both the transaction quantity and price shall meet the following conditions: (i) Upon closure of the auction, the transaction system shall rank the declarations by price, then rank the same-priced declarations by time. (ii) If the total declared quantity is no more than the total issuance quantity, all the transactions will be concluded at declared quantity at reserve price. (iii) If the total declared quantity is above the total issuance quantity, then all the transactions will be concluded at issuance quantity at the minimum offer price.

6.4.3 Allocation Correction

The covered enterprises under Guangdong ETS regulation vary from each other in business scale and industry categorization. To safeguard fairness and efficiency of allowance allocation, the competent department of Guangdong ETS, while stitching to established allocation methodologies and approaches, has laid out the following adjustment mechanism to deal with special situations in real economic operations:

(1) Fall-back correction to benchmarking allocation

Since benchmarking allocation is closely related to companies' production scale, it is logic to base on the historical average output of covered enterprises for benchmarking allocation. However, historical output, which is a reflection of the past, varies from a company's present production, particularly the companies that are more sensitive to the changes in national economy. Thus, an allocation plan with historical output as the major parameter tends to differ greatly from the companies' actual demand for allowances. When the national economy takes a downward trend, the allowances based on historical output will exceed the actual demand, then leads to surplus allowances, which will weaken the ETS effect emissions abatement. On the contrary, there will be a tight allowance supply if the allowances are below the actual demand, which will impose great cost pressure on the companies' normal production. Therefore, a fall-back correction to benchmarking allocation will enable scientific management of allowances in light of the companies' real production activities.

(2) Cement, petrochemical, and iron and steel sectors: set a cap on allowances allocation

Benchmarking allocation is available for fall-back correction, yet such correction may not be unrestrained. It is a feasible option to adjust the upper limit of output in light of designed capacity. If the output is above the capacity, then switch the output-based allowances to the capacity-based allowances, and allow the companies to purchase additional allowances to make up for the deficiency. Such approach, on the one hand, satisfies the development demand of companies, and on the other levels the playing field and controls aggregate emissions of a given industry, and injects vitality into carbon market, which is ought to be accepted by most participants.

(3) Deal with special circumstances

Since the economic practice is always sophisticated, it is natural to see some special circumstances during the pilot period of Guangdong ETS. The citation of some typical circumstances and resolutions are given as follows:

The 2013 allocation is, in principle, based on the verified annual average emissions over 2010–2012 reported by the covered enterprises. But the adoption of historical data may follow some special rules in case of the following exceptional circumstances: (a) If the companies (production lines or generators) are put into operation (or major technical revamp) in any year over 2010–2012, if their normal production fails to reach 12 months in a row in 1 year, then the emissions of next year (12 months) will be defined as the benchmark data. (b) Over 2010–2012, if the companies (production lines or generators) are halting production for over 1 month in any year owing to force majeure, then the emissions data from that year will be put aside. (c) If the companies (production lines or generators) are put into operation in 2013, yet fail to report the emissions data from more than 6 months of stable operation, then they shall abide by the rules that apply to New Entrants in the carbon market. (d) If the companies (production lines or generators) autonomously halt production for over 6 months in any year over 2010–2012, and fail to start up till December 2013, then the allowances for 2013 will be frozen, and subject to reassessment by the competent department after they officially resume operation. (e) If the companies (production lines or generators) are blacklisted for elimination, whether they are closed in a given year over 2010–2012, the allowances are for only valid for that year. If the closure is carried out in advance, their remaining allowances shall be withdrawn.

6.5 Crucial Elements of Guangdong Pilot ETS Allowance Allocation Plan

An overall allowance allocation plan is made up of methodologies, criteria, and distribution cycle, which may implicate detailed aspects of the practical operation. This section, by drawing the experiences from implementation of Guangdong ETS pilot program, sheds light on some crucial elements of the allocation plan.

6.5.1 Objects of Regulation

By drawing the experiences from carrying out the pollutant discharge standards, and other mandatory measures for control over energy consumption, which regulate emissions from companies and gather their emissions data; and the companies are adapt to reporting wholesale emissions data instead of breaking down to the emissions from installations. Therefore, China's seven ETS pilot programs define companies as objects of regulation and sources of data, in contrast to EU ETS and other foreign cap-and-trade systems that regulate installations. Take Guangdong ETS, for instance, it is the covered enterprises that report emissions data and receives allowances, and their business scope is defined via an exclusive Organization Code; in other words, each Organization Code accesses to a registered account. However, such regulation with companies as centerpiece has the following problems:

(1) Inexplicit emitting sources. Each company may have different types and quantity of emitting sources. The companies, despite the same industrial property, may have diverse emitting sources, which is a resistance for benchmarking allocation. Therefore, the special emitting sources shall be handled separately, which inflicts difficulties upon allowance administration. Take cement sector, for instance, some cement plants buy minerals directly, while other have self-run quarries, in this case, the CO_2 emissions from extracting quarries are combined into the aggregate emissions of this cement plant; otherwise, additional allowances for emissions from quarrying shall be distributed. Guangdong ETS chooses the second option.

There is an alternative better for clarifying emitting sources: split a company's different emitting sources into working processes (generators or products), develop a specific allocation criterion for each working process (generator or product), and add these allowances into the sum for the companies.

(2) Changing production activities cause a gap between planned allowances and actual emissions. Any technological modification or product switching is a challenge against the allowance allocation to the companies. This is a non-distinctive issue for the companies with stable raw material or product mix or for the single electricity producers. Yet it is another landscape for petrochemical companies, particularly those small-sized ones. Owing to varied raw material sources and oil demands, petrochemical companies are forced to substitute their large equipment (emitting sources), so the allowances defined before such changes will notably vary from the current actual emissions. Under such circumstance (significant changes on production activities), the Guangdong ETS-regulated companies will revise their emissions monitoring plan within 3 months as of the change takes place, and resubmit the plan to the competent department for recording; then the competent department shall designate experts and verification institutions to conduct on-site verifications to the companies, and then ratify proper adjustments to their previous allowance allocation plan.

6.5.2 *Rediscussion on Responsibility Division for Emissions Reduction*

In view of energy consumption and CO_2 emissions, the entire electricity system could be divided into the production side and demand side. For both Guangdong and the national uniform carbon market, whether electricity producers—generating and supplying electricity to consumers—shall take charge for the emissions from electricity generation is a foremost issue for developing allowance allocation plans.

Through surveys of electricity companies and other stakeholders, we found three dominant opinions about responsibility division for emissions reduction [6]: (a) The emissions that fall into the ETS regulation stem from electricity consumption by both power plants and consumers. Those who hold such viewpoint believe the power plants take sole responsibility for the emissions from their own electricity use; as for the electricity supplied to consumers, any emissions therefrom shall be compensated by the consumers. Anyway, the consumers will turn to other fossil energy for power in absence of electricity. (b) The emissions shall be compensated by the companies in the places where the emissions are generated. Such viewpoint, which features clear-cut division of responsibility and helps implementation of emissions credit transaction, is widely executed by the world leading emissions trading systems, e.g., EU ETS and RGGI. (c) Both electricity production side and demand side shall be covered by the ETS mandatory regulation.

The electricity pricing in China is relatively fixed, the administration of emissions allowances tends to increase the cost of power plants, yet they are confined to pass the additional cost onto consumers. Though Viewpoint A may somewhat resolve such contradiction, it has pronounced limitations: the electricity consumed by power plants is generally 5% of their total electricity output, implying that the consumption side—all sectors that consume electricity—shall fall into strict regulation and restriction of the emissions trading scheme. However, a high-degree dependence on the cutback efforts of the consumption side is from enough to fulfill the overall emissions reduction target; after all, CO_2 is mostly emitted from electricity producers. Moreover, the emissions restriction upon the consumption side is fairly complicated, since a great number of sectors and entities are involved, and they have disparate emissions reduction technologies. Therefore, Viewpoint A is hardly able to restrict the emissions from such leading emitters as fossil fuel-based power plants, which, of course, weakens the ETS effect and encumbers fulfillment of the overall emissions reduction target.

Viewpoint B holds that the electricity consumption side should not fall under mandatory compliance for cutting emissions; instead, if they can generate any cutbacks from emissions reduction efforts, these cutbacks may be traded for earnings in the form of CERs under the framework of CDM, GS (the Gold Standard) of other global voluntary emissions reduction organizations, or the CCERs stated in China's *Interim Measures for the Administration of Voluntary Greenhouse Gas Emissions Reduction Transactions* (NDRC [2012] No. 1668). But the current emissions trading market across China remains inactive with low carbon prices, which is hard to motivate the consumption side to save electricity consumption and reduce emissions.

Viewpoint C argues that there is double counting of CO_2 emissions arising from both electricity production and consumption. Under the ETS framework, the allowances to be issued will exceed the needed amount, which may lead to a misconception that the ETS-regulated emissions are more than actual emissions, but in fact it is only a matter of statistics. Such double counting is easily resolved by providing an explanation or taking off the double-counted portion when keeping statistics of the aggregate emissions. This viewpoint, at the production side, appeals the electricity companies to raise the efficiency of generating units and switch to clean energy; while at the consumption side, it calls the energy-intensive companies, particularly cement plants, steel mills and shopping malls, to adopt advanced technologies and raise electricity-use efficiency. The significance of Viewpoint C lies in the equal emphasis on controlling electricity-induced CO_2 emissions (emissions magnate) from upstream and downstream sectors, which is able to maximize the effect of ETS, complete the task of capping on aggregate emissions and fulfill the regional emissions reduction target.

Guangdong ETS is now regulating CO_2 emissions from both electricity production and consumption sides—a core of Viewpoint C as noted.

6.5.3 Problems in Defining Benchmark Value and Resolutions

The accuracy and integrity of companies' benchmark emissions is of great significance for safeguarding the effectiveness of allowances allocation. Yet, the data accuracy is affected by the following factors:

(1) Unstable production of covered enterprises during reference years

Owing to volatile macro-economy or changing supply–demand patterns of certain products, the companies may generate different amount of emissions every year. In order to mitigate the disparity in annual emissions, the benchmark value shall be the companies' annual average emissions in past 3 years or in several given years.

(2) Companies failing to reach prescribed production duration due to newly started operation or major technical revamp

If a company (production line or generator unit) fails to reach 12-month operation in the current year, then the allowances shall be based on its average emissions of the next full year. For a company (production line or generator unit) that is operated in a year or so, the allowances shall be based on the emissions from a normal production period (12 months in a row). If the duration of normal production is shorter than 1 year, then the allowances may be converted on basis of the emissions during the months of normal production, or in reference to the average historical emissions from the counterpart companies.

(3) Special circumstance during the data-collection period

During the production process within the years for drawing the basic data, any inappropriate production management, changing natural environment or social environment may lead to abnormal production, which will then bring uncertainties upon allowances determination. When deciding the quantity of allowances, full consideration shall be given to the reasons, duration, and impact of abnormal production.

If the abnormal production arises from changes on demand–supply pattern or equity structure, or from technical revamp, all the risks therefrom are due to borne by the companies themselves, so allowance determination shall follow the regular procedure, in other words, there is no special treatment.

Special consideration is given to the following two situations: Any abnormal production arises from force majeure, e.g., typhoon, earthquake, flood, fire, or explosion, and extends over 1 month. Any abnormal production under government mandate (excluding electricity generation dispatching) and extends over 1 month. Under such circumstances, the allowances may be calculated disregard of current year's emissions, or converted based on the emissions from normal production.

6.5.4 Matters Needing Attention When Defining Benchmark Values

In light of the allocation methodologies and benchmarking experiences in the foreign emissions trading systems, the benchmarking value for the same group of products is usually the average value of the ETS-regulated products or the top 10 emitters, or the benchmark value of such products. China is now at the period of economic transition, different regions vary greatly in technological strength, sweeping approach is not recommended. For instance, for some small-sized cement plants that use shaft kilns, they are confined by production scale and production process, their efficiency in cutting CO_2 emissions never catches up those large and advanced companies. For the generating units with different production capacities, uniform benchmark value is, of course, unfair; moreover, some small coal-fired power plants in remote areas, despite of low efficiency, are still a backbone for people's livelihood.

Such a situation calls for fairness. Define benchmark value based on types of production lines or generating units. Take account of the situation of small companies and leave them a buffer zone and gradually integrate into the carbon market. In the meantime, in order to protect the leading companies, the priority shall be given to efficiency, the benchmark value of different production capacity may be defined by different approaches. In order to mobilize the initiative of cement producers, Guangdong ETS exercises differentiated allowance allocation criteria toward different clinker production lines. For the production lines at or above 4000 t/d, the BMi is the unit average emissions from the producers of the same technical group. For the production lines no less than 2000 t/d and less than 4000 t/d, the BMi is the unit average emissions from the leading 75 producers of the same technical group. For the production lines less than 2000 t/d, the BM_i is the unit average emissions from the leading 50 producers of the same technical group. In addition, adjust the range of benchmark value in light of the domestic or global level of the same technologies, so as to guide the companies to step onto a high-efficiency and low-carbon development path.

In sum, the definition of benchmark values shall ensure efficiency and fairness of allowances allocation; in other words, when computing quantity of allowances, both industrial baseline emissions and companies' historical missions shall be taken into account, with an aim to protect the advanced companies and winnow out the outdated ones.

6.5.5 Allowances Correction

The CO_2 emissions from companies vary from time to time amid volatile economy and uneven industry development. As a result, an elaborate allowance allocation plan shall either involve an ex-ante approach that applies to economic ups and downs or fall-back approach that deals with production fluctuations. All the pilot carbon markets in China have established an allowance adjustment mechanism. Nevertheless, with regard to the integrity of an emissions trading scheme, an ex-ante approach that foresees all possibilities is better for ETS steady performance.

In order to ensure stability and continuity of the relevant policies and allocation system, the fall-back approach shall be able to solve the practical problems in line with the established basic principles and allocation methodologies, as well as the guidelines for efficiency and fairness. The fall-back approach is not allowed unless the companies indeed have inadequate allowances, increased production, and higher efficiency.

References

1. Carbon Trading Task Group of CAS Guangzhou Institute of Energy Conversion, *Q&A about Design and Practice of Guangdong Emissions Trading Scheme*, China Environmental Science Press, 2014.
2. JIANG Xiaochuan, *Study on the Initial Distribution System of Carbon Emissions in China - Taking the Allocation Method as the Center*, Jiangxi University of Finance and Economics, 2012.

3. Guangdong Province Development and Reform Commission, *the First Allocation and Work Plan for Carbon Emission Permits* [EB/OL]. http://zwgk.gd.gov.cn/006939756/201401/t20140123_463411.html.
4. Guangdong Province Development and Reform Commission, *the Distribution Plan of Guangdong's Carbon Emission Allowance in 2014* [EB/OL]. http://zwgk.gd.gov.cn/006939756/201504/t20150430_578928.html?keywords=.
5. Guangdong Province Development and Reform Commission, *the Distribution Plan of Guangdong's Carbon Emission Allowance in 2015* [EB/OL] http://www.GDDRC.gov.cn/zwgk/tzgg/zxtz/201507/t20150713_322106.html.
6. LUO Yuejun, LUO Zhigang, ZHAO Daiqing, *Mechanism Study on Carbon Emission Trading of Power Sector* [J]. Environmental Science & Technology, 2014, 37 (6N): 329–333.

Chapter 7
Emissions Monitoring, Reporting and Verification

7.1 Concept of MRV and Foreign and Domestic Studies

Carbon emissions transaction is made up of several links, e.g., emissions Monitoring, Reporting and Verification (MRV), allowances allocation, carbon pricing, transaction platform and transaction rules. MRV is an indispensable component, since data accuracy and truthfulness determines allowances computation, allocation efficiency and fulfillment of emissions reduction target. In aggregate, an appropriate MRV regime is an essential part for smooth operation of an emissions trading scheme [1].

7.1.1 Definition of MRV

MRV regime, which is a process of monitoring, reporting and verifying the GHG emissions data of the ETS covered enterprises, is the backbone for normal operation and healthy development of a carbon market. The emitters (companies) are responsible for monitoring and reporting their GHG emissions, while a third-party verification institution accredited by government undertakes emissions verification.

"Monitoring" refers to measuring, acquiring, analyzing and recording the relevant emissions data (e.g., energy and supplies) of the covered enterprises with a slew of technologies and measures.

The validity, reliability and accuracy of monitoring are the basic principles of MRV regime. Data quality control—the foremost part of MRV—runs through the entire MRV process. The content of monitoring is made up of three parts:

(1) Review and cross check all record sheets, including the invoices or receipts for purchasing energy and supplies, energy and supplies consumption records, product sales invoices, activity data summary sheets, and checklists.

© China Environment Publishing Group Co., Ltd. and Springer Nature Singapore Pte Ltd. 2019

D. Zhao et al., *A Brief Overview of China's ETS Pilots*,
https://doi.org/10.1007/978-981-13-1888-7_7

(2) Further examination of all elements of carbon verification, including the definition of organizational and operational boundaries, identification of emitting sources, collection of activity data, selection of emission factor, application of quantitative methodology, and calculation of emissions.
(3) Conduct horizontal and vertical comparisons of the emissions data from the same industry and from the same-sized companies, and compare carbon intensity within the same industry.

"Reporting" deals with processing, integrating and computing the monitored data, and delivers the final version of monitored facts and data to the competent department in a standardized manner, e.g. spreadsheet or paper document in report template.

"Verification" is carried out by an accredited third-party verification institution. Through document review and on-site surveys to the covered enterprises, this institution validifies the companies' emissions data, produces a verification report, and endorses authenticity and reliability of the data. Verification is an independent process with an aim to validify the reported data and information are substantially accurate, and subject to monitoring as requested.

An integrate MRV regime helps build confidence of the stakeholders in the data, and in the entire emissions trading system.

7.1.2 Significance of MRV

MRV usually manifests itself in the form of government document, which lists out the monitored GHG emissions data of the regulated emitters, reporting and verification rules (including fill-in requirements and technical norms), in an aim to ensure consistency, integrity, transparency and accuracy of the reported data.

"Integrity" indicates the monitoring and reporting shall cover all emitters within the companies' organizational boundary; neither omission nor concealment is allowed. Besides, in addition to the emissions from normally-operating installations, the emissions from switching on or down the installations and from other emergencies are also monitored and reported.

"Transparency" means transparent acquisition of actual emissions and transparent computation of allowances. All the data stored and managed shall be transparent, clear-cut and traceable. All data sources and computation approaches are explainable. There are elaborate historical proofs, records and accessory evidence. Avoid to use of ambiguous data.

"Consistency" refers to consistent monitoring methodologies for different industries, regions and timescales; in other word, there are emissions from a specific installation at different times, and the emissions from the same-type installations owned by different companies within the same period, the monitoring methodologies shall be always uniform. Explicit explanations shall be made for possible inconsistency.

"Accuracy" requires accurate computation of the CO_2 emissions from the companies' production and management activities, and minimization of deviations and uncertainties.

7.1.3 Basic Flow of MRV

Basic flow of MRV mainly consists of two parts: (1) The companies report their emissions, monitoring plan and annual emissions data to the competent department; (2) The competent department or an accredited third-party verifier validates the reported data.

The companies shall produce an emissions monitoring plan which covers all important information about their monitoring system, so as to prove the accuracy of the measurement and reporting about the emitting sources. Such plan is then delivered to the competent department for approval. The monitoring plan is subject to dynamic adjustment, i.e., whenever an significant change happens to the monitoring system, the monitoring plan is placed under continuous revision and re-applied to the competent department for approval and recording, and then the companies shall start monitoring of all regulated emitting installations, and report their annual emissions (in standardized report template) to the competent department.

The competent department reviews the monitoring plan submitted by the companies, and then conducts spot check of the emissions reports in reference to the monitoring plan.

At the authorization of the competent department, an accredited third-party verification institution reviews and validates the monitoring plan and emissions reports submitted by the companies, then produces validation opinions and reports to the competent department.

MRV basic flow is shown the in Fig. 7.1.

7.1.4 Foreign and Domestic MRV Studies

The core content of MRV regime comprises formulation of rules and criteria, definition of verification objects and data quality control. The MRV rules and criteria adopted by different countries usually involve companies' monitoring, quantification and reporting guidelines, verif

The EU ETS regulation is split into three phases where the content of MRV regime changes accordingly [2]. With respect to the scope of monitoring, increasing categories of GHGs are involved by the ETS: Phase 1 and 2 only regulates CO_2 emissions, while Phase 3 regulates CO_2, N_2O and PFCs emissions. With respect to the monitoring procedures, more emphasis is given to the monitoring plan. With respect to the monitoring methodology, a uniform verification system and

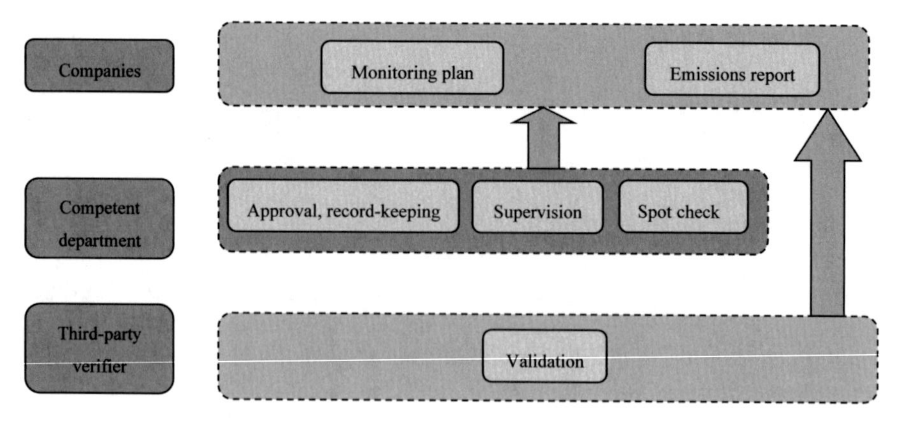

Fig. 7.1 MRV basic flow

accreditation criteria are established: an accredited third-party verifier takes charge of emissions verification, and the competent department validates the emissions reports.

The EU MRV regime is backed up by a series of rules and criteria: 280/2004/EC,[1] 2007/589/EC,[2] Verification of Greenhouse Gas Reporting and Accreditation of Third-Party Verification Institutions [3]. The scope of verification involves five industrial sectors, i.e., energy supply (electricity, heat and steam generation), oil refinery, iron and steel, building materials (cement, lime and glass), pulp and paper. In addition, the accreditation authority is subject to peer assessment in line with ISO/IEC17011:2004.[3] The verification institution is under administration in line with ISO 14065:2007.[4]

As for Tokyo ETS, the MRV regime is backed up by a series of rules and criteria: *Quantification Guidelines for GHG Emissions from Installations*, *Verification Guidelines for GHG Emissions*, *Guidelines for Registration and Application of Third-party Verification Institutions*. There are 1100 commercial installations and 300 factories falling into the scope of verification. In 2008, the Bureau of Environment of Tokyo founded Japan Verified Emissions Reduction (J-VER) which takes the following characteristics:

---

[1]280/2004/EC Decision of the European Parliament and of the Council concerning a mechanism for monitoring Community greenhouse gas emissions and for implementing the Kyoto Protocol.

[2]2007/589/EC Guidelines for the monitoring and reporting of greenhouse gas emissions.

[3]ISO/IEC17011:2004 Conformity assessment—General requirements for accreditation bodies accrediting conformity assessment bodies.

[4]ISO 14065:2007 Greenhouse gases—Requirements for greenhouse gas validation and verification bodies for use in accreditation or other forms of recognition.

(1) A top-down administration model is adopted for developing methodologies and defining benchmark values.
(2) Relevant systems are established in reference to the international norms. For instance, ISO14064-2[5] is consulted for setting up the framework of MRV regime, monitoring and computation rules; ISO14064-3[6] is consulted for drawing up verification guidelines; and ISO14065[7] is consulted for defining the matters for accrediting verification institutions.

California began to ask the regulated emitters to report their annual GHG emissions since 2008. The covered enterprises are not allowed to access to the carbon market unless they have registered via Air Resources Board (ARB) and are ready to take mandatory verification by an accredited third-party verifier. In addition to the six GHGs that are regulated by all leading cap-and-trade mechanisms, California extends its coverage scope to NF_3 and other fluorides. About 350 companies and over 600 installations have joined in Cal ETS, i.e., extending from electricity sector and large industrial installations to fuel distributors in a stepwise manner. Cal ETS has developed emissions computation methodologies for 20 industrial sectors, insists on third-party verification, and organizes strict training and accreditation of such verification institutions. Currently, there are over 40 accredited third-party verification institutions with more than 200 verifiers [4].

Similar to the MRV rules in the EU and California, China's MRV regime also requires submission of annual emissions report, and third-party verification of emissions and companies' compliance. China is now developing a GHG emissions accounting and reporting system for key companies (institutions). It has already defined the reporting threshold, and published the GHG emissions accounting and reporting guidelines for 24 sectors (10 guidelines have become a national norm). It will build an administration system for verification institutions at both national and provincial/municipal levels. The current MRV regime in China's seven pilot carbon markets is briefed as follows:

Shenzhen has built an expert team for establishing GHG verification criteria, developed two universal guidelines, a set of methodologies for two special industrial sectors, and two universal technical specifications. There are also the *GHG Emissions Quantification and Reporting Methodologies for Electricity Sector*, *GHG Emissions Quantification and Reporting Methodologies for Water-supply Sector*, *Notice on Defining the Technical Requirements for Carbon Emissions Verification*, and *Specification on Uniform Adoption of Emissions Factors*. The objects of verification are those legal person-led organizations. In order to safeguard

[5]ISO 14064-2: Greenhouse gases—Part 2: Specification with guidance at the project level for quantification, monitoring and reporting of greenhouse gas emission reductions or removal enhancements.

[6]ISO14064-3 Greenhouse gases—Part 3: Specification with guidance for the validation and verification of greenhouse gas assertions.

[7]ISO14065 Greenhouse gases—Requirements for greenhouse gas validation and verification bodies for use in accreditation or other forms of recognition.

data quality, Development and Reform Commission of Shenzhen Municipality has developed a 'statistics sheet for major energy-intensive equipment in organizations' for cross-checking accuracy of data.

Beijing has published the emissions counting guidelines for six industrial sectors, i.e., heat production and supply, thermal power generation, cement production, petrochemical production, tertiary industry and enterprise industry. For the key emitters (the aggregate direct and indirect emissions from fixed installations are at or above 10,000 tCO_2/year), conduct actual measurements of the activity level and emissions factors in the annual report, and calculate their GHG emissions during the reporting period. For the key installations (the direct emissions are above 10,000 tCO_2/year), conduct separate measurement. The reporting entities shall deliver their previous year' emissions report (in electronic form) to Beijing Municipal Development and Reform Commission (BJ DRC) before each March 31. The key emitters shall submit their previous year' emissions report and verification report (in paper form and sealed) to BJ DRC before each April 30. Conduct dynamic administration of the third-party verification institution, set strict rules for accession and a strict penalty mechanism, and constantly improve professionalism of verifiers.

Tianjin asks the ETS covered enterprises to deliver their next year's emissions monitoring plan to Tianjin Development and Reform Commission (TJ DRC) before each November 30, produce the previous year's emissions report in the first quarter of each year, and deliver both emissions report and verification report produced by a third-party verification institution to TJ DRC before each April 30. Besides, Tianjin has established a mandatory reporting system for key CO_2 emitters, i.e., even if the reporting companies are not qualified to join in the ETS, they are obliged to deliver previous year's emissions report before each April 30, but monitoring plan and verification report are exempted.

Shanghai asks the ETS covered enterprises to deliver their next year's emissions monitoring plan to Shanghai Municipal Development and Reform Commission (SH DRC) before each December 31; both regulated and reporting companies shall deliver their previous year's emissions report before each March 31. A third-party verification institution recording system and verification rules are established, the name list of such institutions are made to the public. The third-party verification institution shall submit the verification report to SH DRC before each April 30.

Chongqing asks the ETS covered enterprises to deliver previous year's emissions report and project-based emissions reduction report to Chongqing Development and Reform Commission (CQ DRC) before each February 20; and upload the annual emissions and reductions to the online reporting system. CQ DRC shall, within 5 business days upon receipt of the written emissions reports, entrusts a third-party verification institution to conduct emissions verification of the concerned companies.

Hubei MRV regime focuses on resolving major issues, i.e., it only regulates the CO_2 emissions from key emitters. The scope of regulation involves direct emissions, emissions from production processes and indirect emissions, but disregards of fugitive emissions. Hubei has established the *Hubei ETS Monitoring, Quantification and Reporting Guidelines*, *Hubei GHG Emissions Verification Guidelines*, and *Recording and Administration Measures for Third-party Verification Institutions*. The ETS covered enterprises are asked to deliver next year's emissions monitoring plan to Hubei Provincial Development and Reform Commission before the last workday of each September, and submit previous year's emissions report before the last workday of each February; the third-party verification institution shall submit the verification report before the last workday of each April.

Guangdong ETS is now regulating the CO_2 emissions from four sectors (electricity, petrochemical, cement and iron and steel). It started counting the emissions from paper, glass, textile, non-ferrous, aviation and ceramics sectors since 2015. Guangdong has come up with the *Carbon Dioxide Emissions Reporting Guidelines for Companies* (for trial), which lists out both general provisions and the specific rules for electricity, petrochemical, cement and iron and steel companies. The MRV rules of Guangdong will be elaborated in Sect. 7.3.

7.2 Current Emissions Metering and Monitoring in Guangdong

In light of the *JIF 1001-1998 General Terms in Metrology and Their definitions*, 'metering' is an activity for unifying measurement units and providing accurate magnitude; 'metrology' is part of measurement, but stricter than common measurement, since it covers the entire area of measurement, and plays a role of guidance, supervision and assurance in measurement.

Metering of energy and supplies involves inspection, measurement and computation of the quantity, quality, performance parameters and characteristic parameters of energy and supplies at each link. Metering is an instrument that provides accurate analysis of energy and material utilization, and forms scientific and unbiased basis for energy and supplies administration [5].

The metering and monitoring of energy and supplies in Guangdong is in line with the *Metrology Law of the People's Republic of China*, *Energy Conservation Law of the People's Republic of China*, *Energy Conservation Regulation of Guangdong Province*, *Measures for the Implementation of the Metrology Law of the People's Republic of China in Guangdong*, and the *General Principle for Equipping and Managing of the Measuring Instrument of Energy in Organization of Energy Using* (GB17167-2006). The major content of metering and monitoring is shown as follows:

(1) In reference to GB17167-2006, all of the energy input-and-output organizations, secondary energy input-and-output organizations and major energy-using equipment shall be equipped with metering instrument. Regular detection and calibration are required to ensure accuracy of such instrument. Gross energy is made up of non-renewable energy (gas, coke, heat, crude oil, natural gas, gas, LPD, refined oil products and biomass energy), and other useful energy through

direct or indirect processing or conversion. The metering instrument consist of electricity meter, water meter, Natural gas flowmeter, loadometer and steam flowmeter [6].

(2) The supplies not only include materials or raw materials, but all the goods that are related to product production, e.g., raw materials, accessories, semi-finished products and finished products. Without uniform criteria, supplies metering mainly relies on the companies' uniform criteria administration system which records supplies warehousing, circulation, utilization and depletion.

(3) A specialized metering administration shall be set up. Experienced metrologists —with professional background and holding at least an intermediate technical title—are invited to take charge of metering administration. Occupational trainings for metrologists and statisticians are held from time to time. All energy and supplies data shall be collectable and traceable.

The companies' emissions verification of Guangdong reveals a fact that large companies are doing a better metering job, particularly electricity companies, they have a full variety of data that are available for cross-checking. Even the companies of different business scope, industrial property and production scale have common ground in energy metering, which is shown as follows:

(1) Weak awareness of metering. Several companies have not yet realized the importance of metering for energy conservation. Owing to incomplete energy metering administration, and halfway implementation of accountability system, the front-line workers regard energy metering as the responsibility of leaders.

(2) Unappropriate metering instrument. The energy metering instrument fails to satisfy the standard and trace the emission sources. The energy input-and-output organizations and secondary energy input-and-output organizations have complete metering equipment, but the major energy-using equipment do not, which is hard to meet MRV's requirements that all monitored data shall be traced back to equipment and units.

(3) Disordered metering administration. Energy and supplies metering is a fairly complicated system that involves companies' supplies purchase, production, consumption and record-keeping. The on-site investigations show that the production department usually keep the metering instrument, the supplies department takes charge of purchasing, the workshops use the instrument, and the statistics department keep metering records, i.e., each department is doing its own way and sometimes passes the buck to the other.

7.3 Compilation of the Guidelines for Reporting of CO_2 Emissions

After Guangdong was designated as one of the first seven areas to carry out the pilot ETS program, Guangdong Government, after making preparations, promulgated the *Interim Measures for Guangdong Carbon Emissions Administration* (No. 197 [2014]

GD Gov) [7] in January 2014, stating explicitly to establish a corporate-level emissions reporting and verification system. In light of the "Interim Measures" and the job requirements for carrying out the ETS program, Guangdong Provincial Development and Reform Commission (GD DRC) organized the compilation of the *Implementation Rules for Guangdong Corporate Emissions Reporting and Verification* (for trial) [8] and the *Carbon Dioxide Emissions Reporting Guidelines for Companies* (for trial) [9]; the latter is split into the separate guidelines for electricity, cement, petrochemical and iron and steel sectors, which take full account of the distinctive production processes of the four sectors as noted, and integrate with the actual levels of production, metering and detection, have become important basis for quantifying the corporate carbon emissions.

The construction experiences of the global carbon markets prove that normal operation and healthy development of a carbon market depends on reliable and accurate corporate-level emissions data, which are based on a standardized emission reporting system. The guidelines for corporate-level emissions reporting have laid a technical basis in this regard, and provided scientific methodologies for emissions quantification, thus vigorously propping up the normal operation of Guangdong carbon market. They play a positive guiding role in the ETS study, companies' transition to low-carbon development path, assessment of energy conservation and low carbon emissions, and implementation of the work for regulating the emissions from key sectors. Besides, they are able to promote low-carbon development of Guangdong-based electricity, cement, petrochemical and iron and steel sectors, and build up their product competitiveness on global market.

The authors of this book joined in compiling the emissions reporting guidelines for electricity, cement, petrochemical and iron and steel sectors of Guangdong, and took charge of the guidelines for cement sector. Their train of thought, compiling principles and relevant policies are introduced in Subsects. 7.3.1 and 7.3.2, which will reveal the key points, difficulties and distinctive creation in the MRV regime for Guangdong ETS.

7.3.1 Compiling Principles

The compilation of the guidelines at the corporate level for reporting of CO_2 emissions is strictly in line with *GB/T 1.1-2009 Directives for Standardization*, and abiding by the basic principles as "correlation, consistency, accuracy, transparency and authenticity".

(1) Correlation: Choose appropriate methodologies for calculating companies' CO_2 emissions. Decomposition calculation is able exhibit the emissions from each production process and their correlations. With feasible technologies and reasonable cost, the companies shall improve monitoring conditions and upgrade quality of the reported data.

(2) Consistency: Use uniform methodologies to define the emissions reporting scope, data gathering, calculation and reporting, and to compare relevant CO_2 emissions information.
(3) Accuracy: Calculate CO_2 emissions from the companies' production and administration activities in an accurate manner.
(4) Transparency: The data gathering and calculation process shall be clear-cut and verifiable. There shall be explanations for the calculation methodologies and data sources.
(5) Authenticity: The data provided by the companies shall be authentic; and the content of monitoring plan and emissions report shall truly reflect the actual emissions.

In addition to the above principles, Guangdong ETS research group proposed to add another two principles for compiling the guidelines at the corporate level for reporting of CO_2 emissions, for the purpose of maintaining practicability and creativity of these guidelines.

(6) Sustainability:

The compilation of the guidelines for companies' reporting of CO_2 emissions shall follow the principle of sustainability, which embraces the calculation methodologies at different levels (whether there are equipment-level metering or measured emissions factors). The information modules for emissions reporting covers the companie's basic information, major product information, emitting installations and emitting sources, and production situation, so as to reflect the companies' emissions in an all-round manner, build a linkage with the potential methodologies and allowances allocation mechanism in the national uniform carbon market in the future. The principle of sustainability helps maintain continuity and stability of the calculation methodologies, sustain the emissions information for the national uniform carbon market, and mitigate any adverse impact from changing calculation methodologies or inadequate information modules.

Sustainability also represents that the guidelines shall be applicable to the reporting demands of the companies of varied size or during different period, and the guidelines are effective during both historical emissions reporting period and official reporting period. For large companies that are capable of measuring emissions, the reporting may adopt measured emissions factors and measured calorific value, the emissions data may be traced back to emitting equipment. For small companies incapable of measuring emissions, the reporting may adopt reference emissions factors and default value (calorific value) free from measurement. For the historical emissions reporting period, in case of data vacancy or incompleteness, relevant default value is used. For the official reporting period, a measurement method is recommended after the emissions data are standardized and refined.

(7) Operability:

The compilation of the guidelines for companies' reporting of CO_2 emissions shall follow the principle of operability: the data sources and statistics shall be based on the companies' current monitoring level and metering system; the data sources at different levels shall be provided to enable the companies, at the prerequisite of operability, improve monitoring and metering. Moreover, the guidelines shall involve standardized filing method and template, detailed filling items, and explanations for the filling content, so as to reduce the reporting difficulties, raise the accuracy and standardability of the reports, and facilitate the third-party verification institution to verify and trace back the companies' emissions reports and monitoring plan.

7.3.2 Basic Framework of the Guidline [9]

The *Carbon Dioxide Emissions Reporting Guidelines for Companies* (see No. 16 in References) consists of ten parts: scope, criteria for quotations and references, glossary and definition, principles, reporting scope, emissions computation methodologies, data monitoring and quality control, monitoring plan, emissions report and appendix. The essential components are explained as follows:

(1) Reporting scope

In light of the *Carbon Dioxide Emissions Reporting Guidelines for Companies*, the reporting scope refers to the year of reporting, identification of organizational boundary, identification of CO_2 emissions activities, identification of CO_2 emissions unit and equipment, and selection of data reporting hierarchy.

(a) Definition of organizational boundary:

Definition of organizational boundary, which aims to clarify the legal statistical scope of a company, consists of an accounting report with the legal person as basic unit, and the company's foundation date, production capacity, business scope, equity structure, asset situation, ownership condition, plan area layout plan and organization chart. Furthermore, detailed explanations shall be made in case there is a corporate group legal person, trans-provincial subsidiary, changes upon organizational boundary or business outsourcing,

(b) Definition of CO_2 emissions activities:

From the perspective of technologies and in light of the physical and chemical properties of CO_2 emissions, CO_2 emissions activities consist of direct and indirect emissions activities, as well as special emissions activities.

Direct emissions are generated from fossil fuel combustion, alternative fuel combustion (excluding biomass fuel), carbonate decomposition, organic carbon decomposition, carbon precipitation and flue gas desulfurization. Indirect emissions arise from consumption of purchased electricity and heat. There shall be detailed definitions and explanations of special emissions.

(c) Definition of CO_2 emissions units and facilities:

CO_2 emissions units and facilities refer to the specific physical location where the emissions activities are taking place. Based on organizational boundary and definition of CO_2 emissions activities, we are able to further identify the corresponding emissions equipment owned by the emissions unit, and the energy and carbonaceous supplies used in the installations. Emissions unit is made up of all fixed or mobile units that directly link with CO_2 emissions (Table 7.1).

Deciding on levels of emissions reporting:

Levels of emissions reporting indicate that, under certain conditions, for each type of emissions activity, the company may choose the emissions data at different levels for computation and summarization. The reported data are from the levels of company, emitting unit or equipment. As for the company's emissions reporting and verification system of Guangdong, the default emissions data are assigned to the corporate level; however, if the emitting unit or equipment is available for better metering, and the accuracy of metering instrument is not lower than the

Table 7.1 Exemplary CO_2 emissions scope [9]

CO_2 emissions scope		Exemplary activities	Exemplary emissions unit	Exemplary emission facility
Direct	Stationary combustion	Combustion of solid, liquid and gas fuels occurring on fixed production equipment, or combustion of other alternative fuels or combustible substance fossil carbon, e.g., coal, petroleum, natural gas, gaseline, LNG, gas, solid waste, liquid waste	Cement: clinker calcination Electricity: generating unit Iron and steel: sintering, iron-making Petrochemical: constant pressure reducing device	Cement: decomposing furnace, rotary kiln Electricity: boiler Iron and steel: blast furnace hot-blast stove, blast furnace Petrochemical: heating furnace
	Mobile combustion	Fuel combustion (gaseline or diesel) from transport activities with the transport means (vehicle, ship) owned by the company/unit itself	Proprietary transport means	Vehicle, ship
	Industrial processes	Carbonaceous raw material processing and non-fossil fuel utilization, e.g., cement, ceramics and lime production (there are CO_2 emissions from decarbonation process),	Cement: clinker calcination Iron and steel: lime calcination	Cement: decomposing furnace, rotary kiln Iron and steel: lime kiln Petrochemical: hydrogen

(continued)

Table 7.1 (continued)

CO_2 emissions scope		Exemplary activities	Exemplary emissions unit	Exemplary emission facility
		iron and steel manufacturing (there are CO_2 emissions from iron-making flux decomposition and steel-making carbon reduction)	Petrochemical: catalytic cracking	production plant
Indirect	Consumption of purchased electricity and heat	Consumption of purchased electricity and heat arising from operational process of company or unit.	Cement: raw meal preparation, cement grinding Iron and steel: catalytic cracking Petrochemical: catalytic cracking	All installations that use purchased electricity
Special	Biomass fuel-based emissions	(i) Biomass fuel combustion: impregnated wood chips, dried sludge, wood, non-impregnated wood chips, agricultural wastes, organic waste, textile waste (ii) Partial biomass fuel combustion: waste tire, waste leather	Cement: clinker calcination Electricity: generating unit	Cement: decomposing furnace, rotary kiln Electricity: boiler
	CO_2 transfer	CO_2 is one of the raw materials for producing carbonated drinks, dry ice, fire extinguishing agent, refrigerant, experimental gases, food solvent, chemical solvent, chemical raw material, paper-making	Entire company	–

corporate-level metering, the priority shall be given to the emissions data from the level of emitting unit or equipment.

(2) Methodology of CO_2 emission calculation

The *Guidelines for the Corporate-level Emissions Reporting* introduce the methods and formulas for quantifying direct and indirect CO_2 emissions. Direct emissions stem from fuel combustion and production, and indirect emissions are generated by consumption of purchased electricity and heat. Such calculation relies on Emission Factor Approach and Material Balance Approach. In addition, Calorific Value Approach and Carbon Content Detection Approach are available for calculating the emissions from fuel combustion.

In order to ensure sustainability and operability of emissions reporting, the *Guidelines for the Corporate-level Emissions Reporting* also provide several reference values for detection, e.g., net calorific value and emission factor of all fossil fuels, and carbon content of alternative fuels, in case the direction detection is infeasible. The reference values and detection technics about emission factor are quoted from the world, domestic and provincial levels and from multiple departments (e.g., the national norms, the NDRC-initiated accounting guidelines, international files, the list of regulated GHGs released by Guangdong), and subject to certain adjustment in light of the special situation of Guangdong.

(3) Data monitoring and quality control

In order to standardize data monitoring and quality control, the *Guidelines for the Corporate-level Emissions Reporting* specify the relevant requirements for data monitoring, recording, filing and uncertainty evaluation; and also clarify the technical norms, detection methods, monitoring frequency and requirements for defining data sources, and for preserving necessary evidence like fuel purchase invoice and test reports produced by technical institution.

There may be uncertainties in obtaining activity data and emission factor. Currently, the cement, electricity, iron and steel and petrochemical sectors of Guangdong are not suitable for quantitative evaluation of uncertainties, so the activity data and emission factor are subject to qualitative explanation; and the companies are obliged to explain the measures that they have adopted to reduce such uncertainties.

(4) CO_2 emissions monitoring plan

In oder for the companies to obtain emissions data, standardize emissions data, and abide by the principles of operability, consistency and sustainability, the Guidelines include templates for emissions monitoring, refine the filing items, make detailed explanations for the content for filing, for the understanding of companies and third-party verification institution. The Guidelines ask the companies, when significant change occurs to emissions data, they shall amend their monitoring plan and not lower the monitoring requirements. Moreover, when any major change occurs to the emissions data, the companies shall retain intact internal record for verification by the third-party verification institution (the template for the CO_2 monitoring plan is attached in the appendix).

(5) CO_2 emissions report:

In light of the principles of operability, consistency and sustainability, for the companies to finish their emissions report in an accurate and convenient manner, the Guidelines include the template for the CO_2 emissions report, and detailed content for filing, and offers detailed explanations for the content for filing, so as to lower the difficulties in filing the report, and facilitate verification by the third-party verification institution (the template for the CO_2 emissions report is attached in the appendix).

(6) Appendix

For the purpose of enabling the companies to complete their CO_2 emissions report in an accurate and convenient manner, the Guidelines provide detailed and complete appendix which involves treatment of special situation of organizational boundary, direct emissions from fuel combustion, emission factor of indirect emissions, net calorific value and emission factor of certain alternative fuels, template for CO_2 emissions monitoring plan and report, for the sake of filing and verification by third-party verification institution.

7.4 Verification Rules and Quality

In the entire MRV design of Guangdong ETS, the covered enterprises are main bodies in report, they shall develop a monitoring plan in light of the *Implementation Rules for Guangdong Corporate Emission Reporting and Verification* (For Trial) [8] (no amendments to the monitoring plan in case of no significant change) and a report on previous year's emissions, and report to Guangdong Provincial Development and Reform Commission (GD DRC) via the Emissions Reporting and Verification System, and coordinate with the third-party verification institution to verify the monitoring plan and emissions report. Through the government-initiated bidding for third-party verification institutions, GD DRC will define one institution to conduct CO_2 emissions verification. Such institution carries out verification according to law and in an independent and impartial manner, take responsibility for standardization, authenticity and accuracy of their verification report, and perform the duty of confidentiality as required by law. The development and reform commissions at all levels administration of such institutions. GD DRC take charge for organization, implementation, comprehensive coordination and supervision of the entire province's emissions administration. The municipal people's governments at and above the prefecture level shall guide and support the emissions administration within their administrative area. The municipal development and reform commissions at and above the prefecture level are responsible for organizing verification of the companies' emissions report (Fig. 7.2).

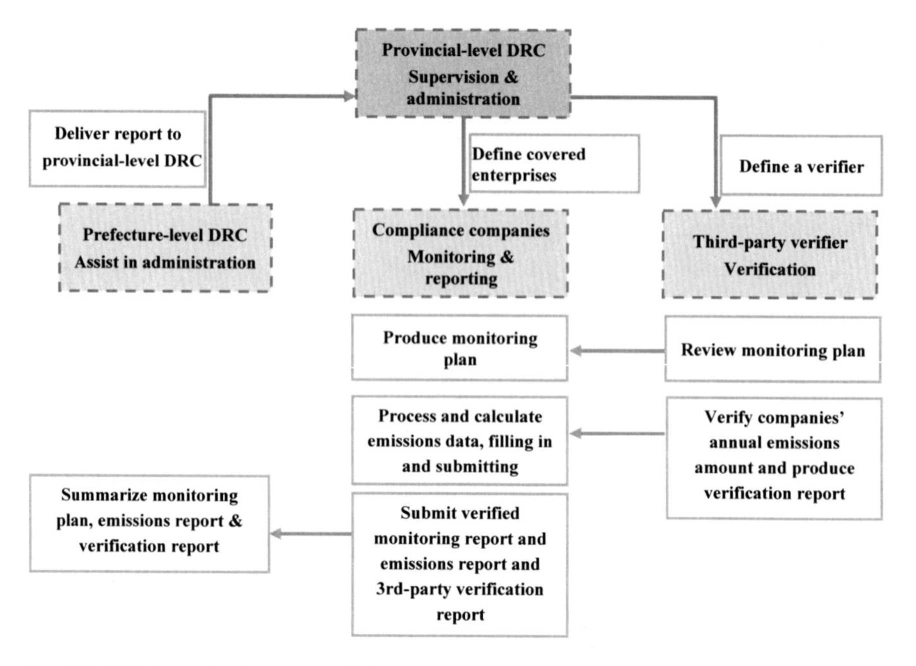

Fig. 7.2 MRV flow in Guangdong ETS

7.4.1 Verification Principles

The *Guidelines for the Corporate-level Emissions Reporting* is a scientific basis for quantifying companies' CO_2 emissions in a standardized and normalized manner; they are also a technical cornerstone for the entire MRV regime by providing a consistent quantification criterion for the emissions from different sectors, different areas or in different time frame, thus enabling transparent calculation of actual emissions and projected emissions. The work of verification, in an aim to check and confirm the integrity and accuracy of emissions data, involves all emitters without exception. Relevant materials for tracing, certification and summarization are backed up during verification.

When carrying out the verification, the third-party verification institution shall follow the principles of independence and impartiality, and moral conduct that is credible, integrate, confidential and prudent. Meanwhile, the verification institution and the company to be verified shall have no conflict of interest for the sake of independent verification.

7.4.2 *Requirements for Verification Competence*

(1) Verification institution

Being the principal party in undertaking the verification, the third-party verification institution shall make sure that all the participating verifiers are capable of fulfilling their assignment and professionally qualified. The verification institution shall establish competency criteria for verification chief, verifier and technical reviewer, particularly for verification chief and technical reviewer, devise the methods for evaluating competence and performance of verifiers regularly, and maintain occupational trainings for verifiers so that they are able to finish the verification successfully within the time limit.

(2) Verifiers

The verifiers shall have enough professionalism and occupational ethics, be skilled in utilizing the principles, procedures and technologies for verification, ensure consistency and systematization of verification, and confirm that the verified evidence is adequate and suitable to draw the verification conclusion. The verification chief, in addition to the abilities of common verifiers, shall be able to preside over an efficient and orderly verification, and lead the verification team to work out the conclusions and produce a verification report. The technical reviewer take charge of reviewing the verification report and internal verification documents, they shall be as competent as the verification chief, and have necessary professional background to judge the integrity and authenticity of the verified information, then accurately assess the verification result.

7.4.3 *Verification Process*

The entire verification process is made up of five parts: contract review and acceptance, verification startup, on-site verification, compilation of verification report and closeout of verification. In order to make sure the verification quality, the third-party verification institution shall, in addition to a verification group, appoint an independent reviewer who conducts final examination of the on-site verification plan, data sampling process, verification findings and conclusions, so as to lower the risks in verification. The process of verification can be represented in the following framework diagram (Fig. 7.3).

The focus of on-site verification is placed on examination of data quality administration system, which is basis and instrument for tracing and confirming the emissions data. The examination of such system shall follow the three steps in the flow chart hereunder (Fig. 7.4):

Data collection focuses on gathering and maintaining the emissions data from major emitting installations, and calibration of detecting devise and metering instrument.

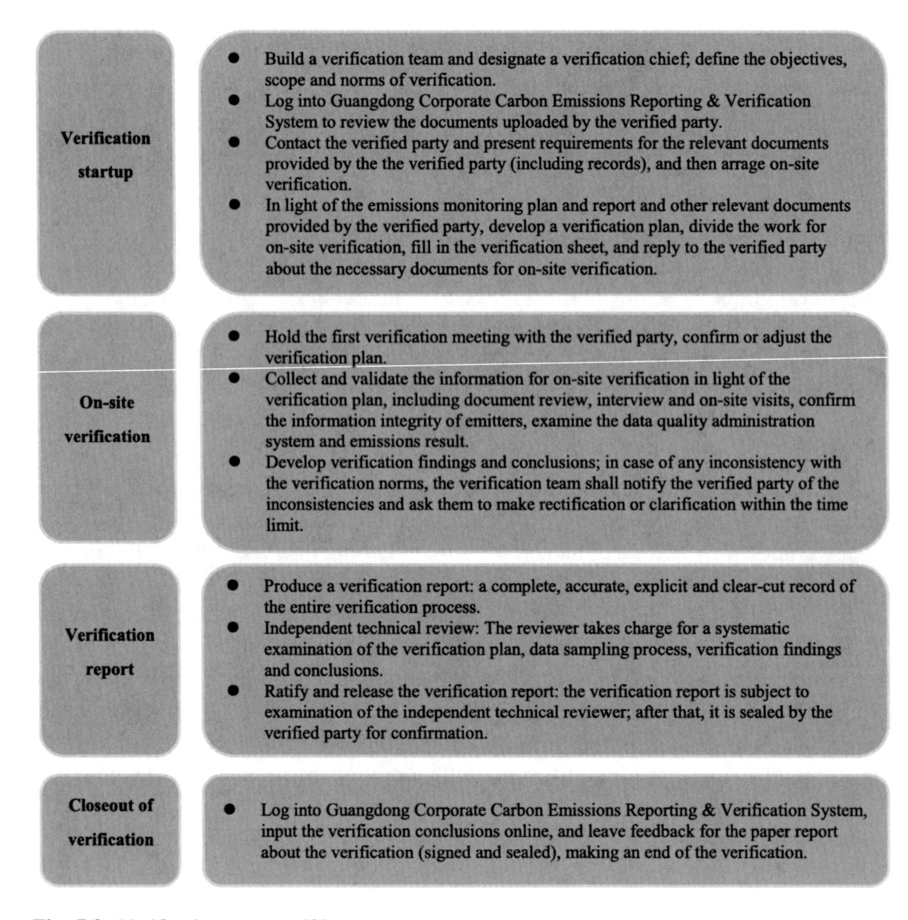

Fig. 7.3 Verification process [8]

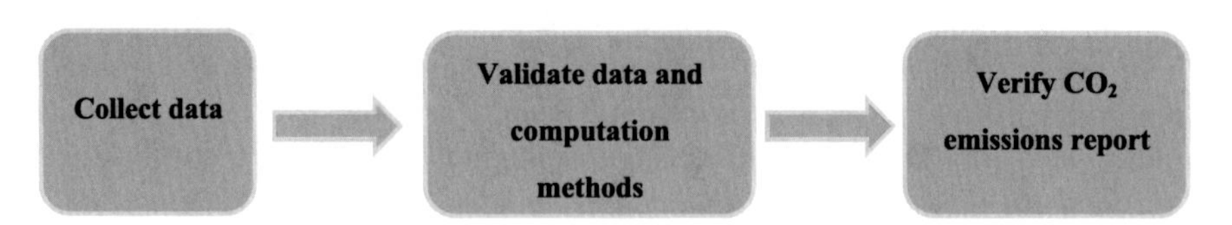

Fig. 7.4 Verification data chain

As for validation of data and computation methods, it is used for examining the relevant records and methods in reference to the activity data, computation methods, emissions factor and emissions amount that are covered in the emissions report, and based on the high-risk links (e.g., key emitters) that are clarified in the sampling plan. The aim of validation is to find out the abnormal value, fluctuations, trends and data deficiency in the emissions report, and confirm if there is any major deviation amid the data provided by the verified party.

(1) Trace the emissions data back to their original records.
(2) Predict the emissions trends through comparison of the historical emissions data.
(3) Evaluate the consistency of the emissions data through comparison of the emissions from similar emitters.
(4) Cross check the emissions data from diverse sources.

The third-party verification institution shall preserve their verification records and relevant papers (in both electronic and paper forms) for ten 10 years since the end of the entire verification work. The materials kept by the verification institution include the materials provided by the verified party, the internal verification papers (important work records and auxiliary materials) and the verification report. Without authorization of the development and reform commission, the verification institution may not disclose any information about the verified party.

7.4.4 Improvement of Verification Quality

Verification quality is affected by the following factors: the competence of verification institution and its verifiers, the competence and coordination of the verified party (regulated companies), and supervision and administration of the competent authority. Therefore, the improvement of verification quality is determined by the competent authority, the verified party, verification institution and its verifiers.

(1) Competent authority

Under the MRV framework, the competent authority takes charge for coordination and supervision of companies' emissions reporting and verification. A properly-designed verification system, selection and supervision of the verification institution are part of the determinants for smooth execution of the verification work. The establishment of a third-party verification system, and fair, objective and transparent selection of a verification institution not only improves verification quality, but foster high-quality verification institutions, which will provide professional services to government and companies.

Guangdong ETS is under administration of Guangdong Provincial Development and Reform Commission (GD DRC) which takes charge for coordination, supervision and administration of the provincial MRV regime, and defines the third-party verification institution by means of bid invitation. GD DRC has announced the strict requirements for verification of in such bidding documents for the verification of the emissions from key companies/institutions and qualifications of verification institutions. Such documents state GD DRC has one-vote veto for qualifications and suitability of verification institution, and present strict requirements: technical and commercial review, understanding of the focal point of the verification work, implementation ideas and plan of verification, basis for work, qualifications, research experiences and requirements for personnel, setting a high threshold for accession of verification institutions. As for administration of verification quality, the verification institution shall, in light of the Notice on the Provisions for Emission Reporting and Verification of Companies (For Trial) [9], The Guidelines for the Corporate-level Emissions Reporting, the Carbon Dioxide Emissions Reporting Guidelines for Companies (For Trial) [10], as well as the CO_2 emissions report, re-examination of verification report, and double check of verification quality.

(2) The verified party

Being the main body that receives verification, the verified party itself (abilities in monitoring and metering emissions and making statistics, and coordination with the verifier) significantly affects verification quality. And their possession of some essential statistics, traceable documents and invoices that concern CO_2 emissions also affects verification quality. In sum, the emissions administration office and energy administration office and staffing of the verified party have significant impact on verification quality.

CO_2 emissions reporting and verification is a wholly new experience for companies, most of them have yet built an emissions reporting and administration system, and not clarified the emissions metering responsibility. At the initial phase of Guangdong ETS, four industrial sectors fall into the ETS regulation, some companies were not at the level playing field for emissions data monitoring, and some companies failed to carry out the work for energy metering and making statistics, and failed to properly record and make archive for monitored data, thus, they were somewhat resistant to CO_2 emissions reporting and verification. Therefore, it is a great stimulus to upgrading verification quality by holding trainings about carbon transaction, emissions reporting and verification, and cultivating talents in carbon and energy management. Along with continuous operation of the ETS, the companies have become highly aware of the importance of the emissions reporting and verification, from passive acceptance to active coordination, which has greatly improved verification quality.

(3) Verification institution

Being the executor of the verification work, the verification institution plays a critical role in ensuring verification quality. The verification institution shall, in light of the verification specifications, carry out the verification work (including document review and on-site verification) according to law and in an independent and impartial manner, take charge for the standardability, authenticity and accuracy of the verification report, perform duty of confidentiality, and bear legal liability. The verification institution shall make sure that all participating verifiers are capable for fulfilling their tasks, particularly the verification chief and technical reviewer, and evaluate comprehensive quality of all participants. Moreover, all verifiers shall attend regular occupational trainings so as to improve verification quality.

(4) Verifiers

The verifiers—direct participants in verification work—play a decisive role in verification quality. Owing to particularity of verification work, and varied capability and understanding of verifiers, there may be disparity in verified emissions that are assigned to individual verifiers with varied common sense and verification scope. Therefore, there shall be standardized and normalized verification procedures, the verification institution shall strengthen occupational trainings and quality improvement of verifiers, so as to mitigate the disparity in verification result arising from individual verifiers.

The verifiers shall bear in mind of the verification principles and procedures and apply them into the practice to achieve a consistent and systematic verification result. They shall have a full understanding of the production processes of the verified party, so that they will be able to distinguish between true information and the fake one when dealing with massive information and data, and then cross check the relevant information. Moreover, they shall be able to examine and analyze the data systematically, giving priority to important issues, particularly the key emitting sources that are exposed to major deviations.

7.5 Dissection of Key Issues and Suggestions

Carbon trading is a newborn event in China, while a standardized and normalized MRV regime, which is foundation for normal operation and sound development of carbon trading, is also new to China. Among one of the first pilot carbon markets in China, Guangdong Province is fully aware of the importance of the MRV regime, so it has poured massive energy into MRV design. When developing the *Guidelines for the Corporate-level Emissions Reporting*, which is recognized as technological basis for the MRV regime, Guangdong had paid attention to lots of practical issues.

7.5.1 *Key Issues in Developing the* Guidelines

(1) Reasonable definition of companies' organizational boundary

Guangdong Province is home to a large number of companies, yet their organizational boundary is fairly complicated, e.g., different companies share the same energy sources, inexplicit property rights, non-independent corporate enterprise, conglomerate and cross-regional enterprises, which has caused great difficulties in defining their organizational boundary. As such, when developing the *Guidelines for the Corporate-level Emissions Reporting*, the authors consulted and surveyed lots of authorities and channels to improve their definition of companies' organizational boundary, and to fix a scientific statistical boundary for companies.

(2) Taking account of the differences in emissions monitoring, metering and detecting among companies

Most companies have yet built a carbon emissions data monitoring system, nor clarified the duties of emissions metering. They are not on a level playing field of emissions data monitoring, e.g., some of them have failed in metering and keeping statistics for energy or failed in recording and archiving the monitoring data. Under such difficult situation, the authors have taken full account of the differences in emissions data monitoring, metering and detecting among companies when developing the *Guidelines*, and they designed varied data hierarchies, defined emissions factor, net calorific value of fuels and other relevant default values, so as to guide the companies to finish the emissions report. The default values are firstly based on the reference values announced by NDRC, NBS and other competent authorities.

(3) Effect implementation of the *Guidelines*

The companies in Guangdong actually have tremendous differences. It is a core mission to develop applicable calculation methodologies during compilation of the *Guidelines*. As such, the authors organized extensive surveys of relevant companies, industry associations and research institutions, held discussions to draw opinions and suggestions from companies and industry experts on the framework of the *Guidelines* and calculation methodologies. Moreover, the authors invited some companies with different production scale and metering strength to produce trial report on emissions and try out the *Guidelines*, and then draw the feedbacks from companies to amend and improve the *Guidelines*. The text of the *Guidelines* is published via official websites, emails, symposiums, seminars and trial operation to solicit the opinions from the electricity, cement, iron and steel and petrochemical companies, research institutions and industry associations for further improvement. The *Guidelines* were finally applied by the electricity, cement, iron and steel, petrochemical companies in producing historical emissions reports in 2013.

7.5.2 Key Issues in Emissions Verification

The MRV regime of Guangdong has been constantly improved. Local companies have paid more attention to emissions reporting and verification, and most of them have shifted stance toward active collaboration with the verification institutions. Through deliberate cultivation and regulation, local verification institutions and verifiers have become more competent. Yet some problems remain in existence:

(1) MRV policies and regulations and administration remain weak

Guangdong has so far accumulated some experiences in MRV construction and operation, but the relevant policies and regulations and administration remain somewhat defective, the technical supporting system needs to be further refined and improved, and the capacity building and financial support in relevant fields shall be strengthened.

(2) Lack of a perfect emissions data monitoring system, levels of data monitoring and metering vary greatly

Among Guangdong-based companies, the current emissions monitoring equipment are built in line with the existing energy conservation and emissions reduction system. Despite of online monitoring in some companies, their focus is placed on such conventional pollutants as dust, nitrous oxide and SO_2, which is to satisfy the general environmental protection standards, instead of the specific parameters (fuel element, calorific value and carbon content) marking GHG emissions. The companies shall strengthen their awareness in energy metering and statistics keeping, and make up for the deficiency in inept recording and archiving of monitored data which may lead to missing of tracing materials and affect the verification result.

(3) Lack of functional divisions and professionals

Without functional divisions or professionals, most companies are incompetent in organizing professionals in quality control and supervision of emissions monitoring, metering, statistics making, data processing, reporting and verification, and in division of responsibilities. It is a common practice for the companies to transfer personnel temporarily from other departments to take charge of emissions reporting and verification, yet they lack basic emissions knowledge, which worsens quality of emissions report and increases difficulties in verification. Therefore, Guangdong Provincial Development and Reform Commission have input massive manpower and material resource into building the carbon transaction skills of companies, organized trainings on emissions reporting and verification, seminars joined by companies' representatives, on-site surveys to companies, on-site guidance of producing emissions reports, invited industry experts to provide professional guidance, helped companies cultivate talents and enhance their carbon occupational skills.

(4) Lack of carbon asset management division and professionals

In Guangdong carbon market, only a small number of companies have set up a specialized carbon asset management division to make statistics of the company's own emissions and evaluate the company's potential asset value. Such weak awareness of carbon asset directly impacts the activity of carbon market. Although the situation is constantly improved, it still needs to be further strengthened.

(5) Administration and cultivation of verification institutions shall be further improved

Although Guangdong has established the admittance criteria for the verification institutions and exercised strict supervision and administration, more work shall be done to further improve their administration and cultivation. For the purpose of improving verification quality, monitoring and management and verification institutions shall be strengthened, the spot check of verification institutions shall be strengthened to assess their verification level. There shall be a system for archiving credit of verification institutions, which will include their wrong doings (receiving bribes from the verified party or any major deviations in the verification result) into the archives. Any institution with a bad record for dishonest behavior shall be disqualified and exposed to the public.

(6) The qualifications and professional skills of verifiers shall be further improved

Guangdong has established strict admittance criteria for the verification institutions, but lack of specific requirements for qualifications of verifiers. The verifiers vary greatly in their competence, which may affect the qualification result. The verification institutions shall strengthen professional trainings and management of their verifiers to prevent from any deviated verification arising from verifiers' individual differences. Moreover, the verification institutions shall improve the professional skills and credit of verifiers, so that they will not muddle through their work; establish an accreditation system that not only benefits standardized management of the institutions, but raises work efficiency and sustains the work of emissions verification. The verification quality will be improved through sustainable supervision of verification institutions and improvement of verifiers' quality.

(7) Tight schedule for verification

The emissions verification of Guangdong concentrates before the deadline of companies' compliance, which exerts high pressure upon both arrangement of verifiers and schedule for on-site verification, which will leave major impact on the verification quality. It is better for the competent authority to shift the emissions reporting to an earlier date, so the third-party verification institution may start verification as early as possible. Besides, the competent authority may call up the verification institutions to report their working progress and the problems before the deadline of the verification, which helps enure smooth completion of the verification.

References

1. Kedi Jing. Mechanism Design of China Carbon Market and International Comparative Study [D], Nankai University, 2014.
2. Liyuan Zhang. Study on EU Emissions Trading Scheme and Inspirations [D], Nanjing Agricultural University, 2010.
3. Lijun Guo, Kai Meng. Design and Application of the MRV Regime in Carbon Trading [J], China Opening Herald, 2013, (3): 109–110.
4. Introduction to north American carbon emissions trading systems. http://wenku.baidu.com/link?url=z-5ChdcawRzw7r6HcUY_THGA9.
5. Yingjun Wang, Reviewing the Past and Looking into the Future: Energy Metering Remains a Long-term and Arduous Task [J], Industrial Measurement, 2012, (Z1): 1–2.
6. Ke Li, Energy Metering from the Perspective of Corporate-level Energy Conservation and Emissions Reduction [J], Energy Conservation Technology, 2010, (04).

7. Interim Measures for Guangdong Carbon Emissions Administration. Guangdong Provincial Development and Reform Commission, 2013.
8. Implementation Rules for Guangdong Corporate Carbon Emission Reporting and Verification (for trial). Guangdong Provincial Development and Reform Commission, 2014.
9. Carbon Dioxide Emissions Reporting Guidelines for Companies in Guangdong Province (for trial). Guangdong Provincial Development and Reform Commission, 2014.
10. Rules for Corporate Emissions Verification in Guangdong Province (for trial). Guangdong Provincial Development and Reform Commission, 2014.

Chapter 8
Development Track and Policy Context of Guangdong Carbon Market

Since implementing the reform and opening-up policy in 1978, China has been making considerable progress in building a socialist economy with Chinese characteristics, and exhibiting the intrinsic value of land, labor force, and capital which function as production factors in value creation and distribution. While the Chinese economic volume keeps expanding the resources and environment, energy and carbon emissions credits have become scarce elements that have received attention of policymakers. As such, establishing a market mechanism with energy consumption and carbon emissions as basic elements has become an especially important task. The *Decision of the CCCPC on Some Major Issues Concerning Comprehensively Deepening the Reform* calls for "implementing sound compensation systems for use of resources and for damage to the ecological environment", and presents that "we will implement a trading system for energy conservation, carbon emission, waste discharge, and water usage rights.[1]

In a stance to respond to the call of the State, Guangdong Province dared to be a pacesetter in exploring control of GHG emissions via market mechanism. In 2013, Guangdong launched the provincial-level pilot carbon market which has maintained smooth operation for 4 years, accumulated rich carbon trading experiences for the nation, and blazed a new trail in combining government administration with market mechanism. In contrast to other pilot carbon markets, Guangdong carbon emissions trading scheme, after being put into online operation, has kept enhancing market mobility, improving the multitier carbon trading market (focusing on paid allocation of allowances), and has made several new breakthroughs, e.g., the first pilot carbon market that realized a more than 1 billion yuan transaction amount; the first

[1]Adopted at the Third Plenary Session of the 18th Central Committee of the Communist Party of China on November 12, 2013.

© China Environment Publishing Group Co., Ltd. and Springer Nature Singapore Pte Ltd. 2019

D. Zhao et al., *A Brief Overview of China's ETS Pilots*,
https://doi.org/10.1007/978-981-13-1888-7_8

pilot carbon market that finished forward transaction of allowances; the first pilot carbon market that attempted an online platform for green financial services; and the first pilot carbon market that has sophisticated experiences in paid allocation of allowances.[2]

It should be mentioned that the development of Guangdong carbon market has undergone ups and downs, rather than a plain sailing. There have been disputes about market positioning, function and effect, and the relevant policies were subject to partial adjustment. Despite all this, Guangdong insists on discovering problems while designing and fostering the pilot carbon market, drawing up experiences and lessons, and then presenting pragmatic solutions, and ultimately forming a basically complete regulatory framework for the carbon market. While sustaining steady operation of the market, Guangdong has made innovative attempts in carbon financing which was highly praised by NDRC. Some innovative policies of Guangdong were even included in the construction plan for a nationwide carbon market; so to speak, an incoming nationwide market is somewhat a reflection of Guangdong market. Being witnesses to the growth of Guangdong carbon market, we believe that a truly mature carbon market shall at least have two kinds of roles: First, discovering prices and regulating supply-demand pattern, enabling low-carbon technologies more commercially feasible. For those companies that are active in emissions reduction, we should grant them with incentives rather than government rewards, which will reduce their dependence on government. Second, efforts should be made to build financial service companies, control compliance cost with the acceptable range of companies, reduce risks, lower the emissions reduction cost of the entire society, innovating financial products, let financial agents join in allocation of carbon market elements, enable carbon assets to be allocated to the sectors, and companies with higher output and better economic efficiency.

In fact, through review of the relevant policies for Guangdong carbon market, since it was opened in 2013, it is obvious that the competent authority of Guangdong carbon market acts within this large logic framework, abide by market rules, optimize relevant policies in a prudent manner, and made sure smooth operation of Guangdong pilot market. At this point that a nationwide carbon market is about to be built, we are trying to, from the perspective of researchers, sort out the development of Guangdong carbon market, analyze Guangdong primary and secondary carbon markets, and put forward suggestions.

---

[2]Guangdong carbon market won total transaction amount above 1 billion yuan, becoming an inaugurator in several aspects across China http://economy.southcn.com/e/2016-05/13/content_147636109.htm.

8.1 Carbon Market Elements

A complete set of carbon market elements consists of trading actors, trading products, and reward and punishment mechanism. This section, with focus placed on certain key elements, will introduce the changes on the relevant policies about Guangdong carbon market and the logic behind them [1].

8.1.1 Trading Actors

Market exchange is an inevitable outcome when social productivity has developed to a certain historic stage, and when carbon emissions credits become a scarce resource and materialized as private rights and interests, a marketplace for exchanging emissions credits comes into being. In a complete carbon trading market, trading actors are made up of the regulated companies, investors, speculators, carbon asset management institutions, and financial institutions.

(1) Regulated companies

Regulated companies are key suppliers and buyers of carbon emissions allowances, their performance directly impacts the vitality of the carbon market. The regulated companies in carbon market are divided into two categories: (i) The companies whose actual emissions are above their granted allowances, they are forced to buy additional allowances to fulfill their emissions reduction obligation. (ii) The companies with actual emissions lower than their granted allowances, and they can sell the additional allowances at their own discretion. Through investigations, we have found out that in case of loose allowances supply, there will be sparse transaction demand in the secondary market. Such situation is associated with the industry properties and ownership of Guangdong-based regulated companies, i.e., there is inadequate effect demand in carbon market. At the preliminary stage of Guangdong ETS, lots of companies had excessive allowances and did not need to buy additional allowances. Moreover, the ETS remains under adjustment, the companies do not have a clear anticipation for the future transactions, they would store up the allowances rather than selling them out, for fear that they may not buy them back in future. In addition, the administrators with state-owned enterprises usually take a conservative strategy to trade in carbon market because they think that "making no mistake is better than making contribution". Such reasons are weakening the mobility and price discovery function of the carbon market.

(2) Investment institutions and speculators

To introduce more trading actors in the secondary carbon market, China (Guangzhou) Emissions Exchange (CGEE) has designed a more proactive guidance policy by incorporating investment institutions and speculators into carbon trading market, since these trading actors will increase mobility to market transaction, and release a signal for both current and future carbon prices, which will enable the companies to take account of profit and loss when making financial investment decisions about energy conservation and emissions reduction projects. Speculators (including individuals), through analysis of the short-term fluctuations of policies and close watch of the companies' compliance, make "buy low, sell high" decisions. Being rivals of investment and financial institutions, these speculators increases transaction mobility and invigorates the carbon market. Moreover, the investment institutions and speculators are able to make reasonable analysis, predict the low-carbon policies both home and abroad, follow the developments of low-carbon technologies, make objective evaluation of allowances value, adhere to long-term value investment philosophy, and divert funds into the resources element market. In order to absorb these traders, they are allowed to enter the secondary carbon market of Guangdong, and join in the allowances auction during the second compliance period. In terms of trading system, Guangdong has lowered the market threshold and adjusted down formality fees to encourage participation of investors (speculators). Qualified foreign investment institutions are welcome in Guangdong carbon market. Several foreign companies have so far opened a CGEE account. At the end of October 2014, British Petroleum Co. Ltd (BP) opened a CGEE account to become the second foreign investor of Guangdong carbon market. By the end of 2014, there were altogether 19 nonregulatory investment agencies, including two foreign agencies and 45 individual speculators.[3]

(3) Carbon asset management agencies

Carbon asset management agencies, which are different simple investors/speculators, are active participants in the secondary carbon market for providing regulated companies with such services as technical services for energy conservation and emissions reduction, allowances trusteeship and replacement. Their participation in the secondary carbon market is to assist regulated companies in fulfilling remissions reduction obligations at a lower cost and in mitigating risks.

Carbon asset management agencies, through professional carbon market operation, make risk hedging and seek profits by integrating them with other businesses. For the regulated companies that do not hire professional staff to manage their carbon asset, but expect to gain revenue and fulfill obligations, they may pay for the service of allowances trusteeship to a carbon asset management agency. In May 2016, Guangzhou VCarbon Investment Co., Ltd signed an allowances trusteeship contract worth of 3.50 Mt with two subordinate companies of Shenzhen Energy Group Co., Ltd.[4]

[3]Allowances auction revenues to be injected into carbon fund for boosting emissions reduction [EB/OL]. Nanfang Daily [2014-12-11] http://epaper.southcn.com/nfdaily/html/2014-12/11/node_8.htm.

[4]Guangdong carbon trading turnover ranked first in China, a nationwide carbon trading platform is brewing [EB/OL]. Clean Development Mechanism in China (http://cdm-en.ccchina.org.cn/) [2016-7-14].

(4) Financial institutions

Financial institutions, through active participation in green financial activities, provide companies with financing services in energy conservation and emissions reduction to facilitate their carbon transaction. While providing the financing services, the security of mortgage is an essential content for credit risk rating by financial institutions. The emissions allowances issued by the government are a certificate of rights and interests, and an invisible asset with both value and use value, thus conforming to the property of mortgage. With extensive participation of financial institutions, the green financing function of carbon market will be substantially displayed.

The above analysis shows that Guangdong is vigorously constructing a multi-player carbon market with regulated companies playing a leading role, supplemented by investors, speculators and carbon asset management agencies, and backed up by financial institutions. Regulated companies and institutions, new project proponents, other qualified organization, and individuals are allowed to trade the allowances. China (Guangzhou) Emissions Exchange (CGEX) has established a member management rule, involving general members, broker members, and proprietary members (basic and dominant members that trade directly via their own account or agents). Both regulated companies and new project proponents are rightful proprietary members that are exempted from CGEX's ratification. The entry criteria set by CGEX are fairly relaxed, i.e., any financial or investment institutions with net asset no lower than 30 million yuan are accessible.

In addition, in order to guard against any market manipulation, Guangdong has set a cap on the annual allowances held by trading actors: (i) Before Guangdong Provincial Development and Reform Commission verifies their annual actual emissions, no trading actor may transfer more than 50% of their free allowances for the current year to their registered CGEX trading account for transaction. (ii) The allowances held by investment institutions and individuals may not exceed 3 Mt.

8.1.2 Trading Products

The products traded in Guangdong carbon market are Guangdong emissions allowances (GDEA) and Chinese Certified Emission Reduction (CCER). The competent authority of Guangdong carbon market has been considering linking the Points System (associated with the Generalized System of Preference, GSP) with carbon transaction, so as to create a landscape where all social members are joining in emissions reduction. Through cooperation with financial institutions, Guangdong carbon market is making constant innovations, i.e., transforming the traded carbon products into green financial products, with an aim to provide better services to regulated companies.

(1) Guangdong emissions allowances

Guangdong Emissions Allowances (GDEA) are quantified rights to emit CO_2 granted by the competent authority of Guangdong carbon market to trading actors. GDEA is both territory and time bounded: on one hand, GDEA flows within Guangdong, and the interregional distribution may not be included into local carbon intensity, which somewhat reduced local government intervention with companies' allowances trading.

On the other, GDEA is allocated annually, and not confined to a certain period; it can be stored, but not loaned, which restricts allowances speculation and stabilizes the allowances market. In addition, in order to prevent from any adverse impact from production fluctuations, the allowances may be pre-allocated to the regulated industrial sectors based on benchmarking methodology, and then subject to ex post adjustment, which helps relieve the situation of either overly tight or overly loose allowances supply arising from fluctuating production; but such practice is adverse for regulated companies to form clear anticipation of allowances quantity, which is bad for invigorating carbon market.

(2) Chinese Certified Emission Reduction

Unlike GDEA, Chinese Certified Emission Reduction (CCER) is defined by the *Interim Measures for the Administration of Greenhouse Gas Voluntary Emission Reduction* issued by the NDRC, put on file by NDRC and registered via the National Carbon Trading Registry. In order to trigger local nonregulated companies to save energy consumption and cut emissions, the *Interim Measures for Carbon Emissions Administration in Guangdong Province* states that "the CCERs for offset may not be over 10% of the concerned company's last year's actual carbon emissions, and over 70% of the CCERs shall arise from the GHG VER projects within Guangdong" (Article 19). Regarding the policies on CCER offset, Guangdong is still the most open and transparent pilot carbon market in China. In the compliance year of 2015, Guangdong carbon market used about 4.20 Mt CCERs for offsetting actual emissions; such quantity of CCERs ranked top among all pilot carbon market. These CCERs, which mainly came from the offset projects in the western provinces like Guizhou, Inner Mongolia, Yunnan, and Qinghai, vigorously promoted the regional coordinated development and made due contribution to China's two markets (domestic and foreign markets).

In order to mobilize the entire society to join in the low-carbon drive, Guangdong has made explorations in building a carbon GSP[5] mechanism that involves emissions cutbacks in both everyday life and work. In addition to commercial incentives and encouraging policies at the initial phase, Guangdong took the lead in establishing a positive guidance mechanism that integrates GSP with CCER, which is named "PHCER".[6] While considering market maturity and CCER

[5]GPS: generalized system of preference.

[6]The letters "P" and "H" in "PHCER" are two initials of Chinese Pinyin, respectively representing "generalized" (P) and "preferences" (H).

offsets, the PHCER only involved a small number of highly appropriated participants at the preliminary stage, with focus placed on innovation.[7]

8.2 Allowances Allocation and Transaction in Primary Market

8.2.1 Logic Behind Market Regulation

The commodities in the primary carbon market of Guangdong are the emissions allowances granted by the government to regulated companies, including both free and paid allowances. Paid allocation of allowances is a distinctive option of Guangdong with an aim to build a primary carbon market, and enhance the awareness of the compensated use of emissions allowances. But it should be noted that companies are main body of a socio-economy, what they are seeking for is development. For that reason, the foremost task of the primary carbon market is to price carbon without inflicting too much burden upon companies. With respect to market design, when formulating the policies for paid allocation of allowances, the following factors shall be taken into account:

- Arouse companies' awareness of compensated use of emissions allowances: Natural resources and environmental space are limited and compensable. Under the prerequisite of limited resource environmental carrying capacity, companies shall use emissions space more efficiently, otherwise, they have to bear economic cost. Paid allocation of allowances is, in fact, inserting carbon prices into companies' decision-making and operation, so as to force the decision makers to care about energy conservation and emissions reduction and pay attention to environmental protection. Based on such consideration, Guangdong Province, when distribution emissions allowances for the first time in 2013, asked all regulated companies to purchase 3% of total verified allowances in the first place, and then apply for the remaining 97% free allowances. By doing so, most regulated companies have accepted the concept of compensated use of allowances, become more active in environmental protection, and keenly felt the pressure from energy conservation and emissions reduction. They expressed that the extensive development should be put aside.
- The percentage of paid allowances shall reflect industrial differences: For different sectors, the potentials or cost for emissions reduction are varied. While considering the correlation between industrial sectors, and the pressure for sustaining GDP growth and creating job opportunities, there must be a soft transition; specifically, when exercising paid allocation of allowances, different sectors need to buy varied amount of allowances. For electricity sector, the

[7]http://www.tanpaifang.com/CCER/201701/1958337.html.

percentage of paid allowances shall increase from 3 to 5%, such percentage for other sectors will remain generally unchanged.

- The prices for purchasing the allowances may not cause too much burden upon companies: In case a carbon market exerts any negative impact on the local economy as a whole, or heavily weighs on most companies, then the carbon prices shall be reconsidered. For instance, in the compliance year of 2013, Guangdong Government set the floor auction prices for emissions allowances at 60 yuan/t, in an aim to seek market-based carbon pricing. However, during the five rounds of auctions in 2014, such floor price at 60 yuan/t was above market expectation, thus failing to reach the objective of mark pricing. The companies said that such prices would occupy a great portion of their cash flow.
- The paid allocation of allowances shall boost up the market expectations: The primary carbon market that deals with scarce allowances and show slow ox to go shall incentivize the companies to pay for these allowances. The competent authority of Guangdong carbon market has come up with the policy for tiered carbon pricing. From September 2014 to June 2015, one auction would be held in each quarter in principle, with prices stepping up from 25, 30 and 35 to 40 yuan/t. Such regulation intends to foster an upward expectation for carbon pricing.
- The primary market shall link with the secondary market: Although the tiered carbon pricing was adopted in 2014, the prices in the primary market were above the prices in the secondary market. At the later phase of auction, the companies preferred buying allowances at the secondary market. Such a situation proved that the primary and secondary markets shall link with each other, any price inversion is temporary, and adverse for sales of government reserve allowances. In 2015, Guangdong brought into a pricing mechanism that links with the secondary market, i.e., define the 80% of the weighted average transaction price during the 3 calendar months before the date of bidding announcement as the minimum valid price; such policy forces the companies to focus on both primary and secondary markets.
- The Catfish Effect shall be brought into subscription of allowances, investment institutions are encouraged to join in the primary market auctions, so as to vitalize market and increase value of allowances. On September 21, 2016, China (Guangzhou) Emissions Exchange held the first bidding for allowances, five investment (financial) institutions took part in subscription, their subscribed volume (510,000 t) accounted for almost 60% of the total subscribed volume, far more than the total subscribed volume of the four regulated industrial sectors [2]. The total subscribed volume was more than the total paid allowances, and all of these subscriptions were valid. Besides, it was the first time for the financial institution to join in the bidding, i.e., there were altogether two investment institutions and one financial institution that won the bid for 35.76% of the total auctioned allowances, their allowances were only next to electricity sector [3]. The reasons for these institutions to join in the bidding: (i) The auction reserve price was fairly low, only 62.3% of the closing price of the secondary market a day ago, there was a strong expectation for appreciation of

allowances. (2) The participants were holding optimistic attitude toward nationwide and Guangdong carbon markets. Guangdong allowances were of fairly high investment value. The participation of investment institutions reflected that Guangdong carbon market has entered into a new stage. The attention and practice of financial institutions will further optimize the major structure of Guangdong carbon market and improve the function of price formation mechanism and bidding mechanism in paid allocation of allowances.

The above factors shall be taken into account while making adjustments to the carbon market. Through a follow-up analysis of Guangdong carbon-related policies, we have found that Guangdong has made bold explorations in paid allowances distributions, conducted promote evaluation, and achieved a favorable result.

8.2.2 Transactions in Primary Market

Guangdong launched the first auction for paid allowances on December 16, 2013, and cumulatively traded 25.84 Mt of allowances with gross turnover of 997 million yuan by March 29, 2016. The primary and secondary markets traded at 15.66 and 10.18 Mt, respectively; the traded CCER reached 1.39 Mt. In the compliance year of 2013, Guangdong organized five rounds of auctions which traded around 11.12 Mt with turnover of 667.40 million yuan; in the compliance year of 2014, Guangdong organized two rounds of auctions which traded around 2.70 Mt with turnover of 73.04 million yuan [4]. The details about these auctions are given in Table 8.1.

Table 8.1 shows that in the first auctions in 2013 and 2014 (on December 16, 2013 and September 26, 2014), all of the allowances to be auctioned were sold out, and the number of bidders far exceeded the final bid winners, marking that the allowances supply was short of demand. In the first auction in 2014, the trading price was even higher than the floor price. As of February 2015, despite the first auction of each year, in all the remaining auctions where the trading volume was no exceptionally lower than the total issued volume. Under such circumstance of oversupply, Guangdong allowances were generally traded at the floor price, and all bidders achieved what they wanted for.

Regarding the sectors that joined in the auctions, electricity sector ranked first with the largest trading volume; i.e., it had altogether bidded for 8.91 Mt of allowances, accounting for 64.48% of the total auctioned volume, followed by cement sector, investment institutions, iron and steel sector, and petrochemical sector. As for the turnover, electricity sector again ranked the first place a turnover of 489.49 million yuan (66.11% of total turnover), followed by cement sector, iron and steel sector, petrochemical sector and investment institutions (see Fig. 8.1; Table 8.2).

Table 8.1 Auctions for paid allowances in Guangdong Province

Auction date	Total issued volume (t)	Traded volume (t)	Strike price/ Floor price (yuan/t)	Number of bidders	Number of bid winners
December 16, 2013	3,000,000	3,000,000	60/60	56	28
January 6, 2014	5,000,000	3,892,761	60/60	46	46
February 28, 2014	2,000,000	1,130,557	60/60	24	24
April 3, April 17, May 5, 2014	3,600,000	1,737,151	60/60	80	80
June 25, 2014	1,865,000	1,362,870	60/60	46	46
September 26, 2014	2,000,000	2,000,000	26/25	33	19
December 22, 2014	1,000,000	701,442	30/30	12	12

Source China (Guangzhou) Emissions Exchange. http://www.cnemission.com/article/hqxx/

8.3 Transactions in Secondary Market

8.3.1 Overview

Guangdong secondary carbon market was officially launched on December 19, 2013, and cumulatively traded 20.59 Mt of allowances with total turnover of 918 million yuan by July 27, 2015, accounting for 36.12% of the total trading volume and 45.28% of the total turnover of the seven pilot markets [5].

Four models of transaction are approved by Guangdong secondary carbon market: choosing listing candidates, one-way bidding, negotiating transfer and listed bidding. So far it is choosing listing candidates and negotiating transfer are adopted.

In reference to the statistics provided by China (Guangzhou) Emissions Exchange [6], in December 2013 when the secondary carbon market made its debut, it traded 120,100 ton of allowances, after that the transaction turned to slack. During June and October 2014 (within the compliance year of 2013), the allowances trading demonstrated blowout growth, with combined trading volume accounting for about 87% of the total of the compliance year. The electricity and cement companies were the most active traders. After September 2014, the transactions again fell into gloom.

Through review of the trading volume, we found out the transactions remained the most active from July 8 to 15, 2014, the daily average trading volume even exceeded 75,000 ton, and the date of July 15 happened to be the last day of the compliance period of 2013. Thus, it can be seen that during the phase when the ETS was just launched, the secondary carbon market was mainly motivated by the

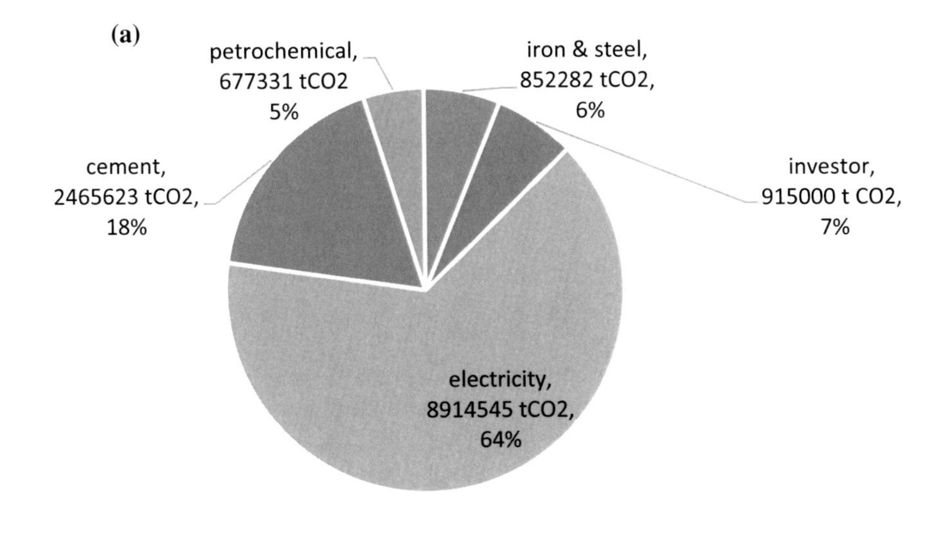

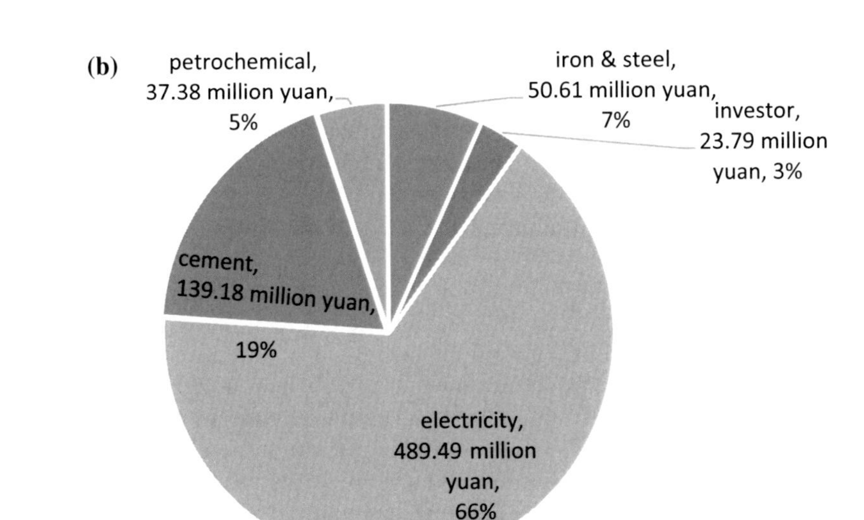

Fig. 8.1 Guangdong allowances trading volume (**a**) and turnover (**b**) by industrial sector

allowances administration system. While the compliance deadline is approaching, and successive execution of the policies for opening the market to institutional and individual investors, Guangdong secondary carbon market has remained robust

Table 8.2 Guangdong allowances auctions in 2014

Bidder sector	Number of bidder winner	Percentage of bid winners (%)
Electricity	65	34.76
Cement	57	30.48
Iron and steel	51	27.27
Petrochemical	10	5.35
Investment institution	4	2.14

before the end of the compliance period, with demand for allowances increasing continuously.

As for the allowances prices, the prices remained in range of 50–70 yuan/tCO_2 by the end of the 2013 compliance period amid moderate fluctuations. In mid-August 2014, Guangdong Provincial Development and Reform Commission released the *Allocation Plan for Guangdong's Carbon Emissions Allowances in 2014* which adjusted the allocation plan for paid allowances. As result of such new polices (e.g., the auction floor prices were vigorously adjusted down), the allowances prices at the secondary market plummeted, and even dropped below the floor price (25 yuan/t) in the first auction in 2014, after that the prices had remained in the range of 12–25 yuan/t.

In terms of the traders, China (Guangzhou) Emissions Exchange (CGEX) have more than 300 members as of end of February 2015, they are made up of regulated companies, new project proponents, institutional and individual investors, service providers and strategic partners. There are 92 members (one-third of the total) joining in the secondary market. Among the total members, there are 53 regulated companies, nine institutional investors and 30 individual investors; cement plants (21) and electricity companies (22) are main traders regulated companies. It can be seen that regulated companies are major players in the secondary market, while institutional/individual investors are emerging forces, they become increasingly active and indispensable in the carbon market.

From the perspective of trading volume, as of February 28 2015, the top three allowances traders had been electricity and cement sectors and investment institutions (see Table 8.3); their combined trading volume held 95.7 and 93.4%, respectively, of the total sales volume and total purchase volume. In terms of the average trading price, the sales prices were leveling off the purchases prices in electricity sector. Institutional/individual investors and petrochemical companies were buying low and selling high, which was especially significant among individual investors, and their purchase volume was above the sales; in other word, their participation in carbon trading was speculative, they were storing up allowances and waiting for the right price to sell. With respect to net purchase and net sales, both electricity and iron and steel companies had more sales than purchase, indicating surplus allowances and smaller pressure for compliance. In contrast, cement companies had more purchase than sales, marking shortage of allowances in the entire cement sector. In light of the study report produced by Guangdong

Table 8.3 Transactions in Guangdong secondary carbon market in 2014

Trading participant	Sales volume (t)	Average sales price (yuan/t)	Purchase volume (t)	Average purchase price (yuan/t)
Investment institution	354,051	49.37	416,141	45.99
Individual investor	12,932	58.64	34,271	35.65
Iron and steel	30,066	36.12	26,468	52.99
Cement	363,617	43.82	473,283	53.18
Electricity	626,844	59.61	422,813	59.61
Petrochemical	17,426	55.18	31,960	42.75

Provincial Cement Industry Association, there used to be massive outdated cement production capacity in Guangdong before 2012, and they had high market occupancy, yet the new production lines were operated constantly since 2010, thus making the cement clinker capacity of regulated companies not fully utilized. As a result, the amount of allowances to be issued in 2013 based on the average cement clinker output over 2010–2012 varied greatly with the actual production of the regulated companies, most of them reported short allowances.

Overall, the allowances traded on the secondary market mainly flow from electricity, iron, and steel sectors to cement sector and institutional investors.

8.3.2 *Market Mobility*

During the compliance period of 2013 (from December 19, 2013 to July 15, 2014), Guangdong secondary carbon market traded 119,1300 ton of allowances with gross revenue of 65.32 million yuan, yet the allowances turnover rate was below 3.4‰. By the end of February 2015, the allowances turnover rate went up to 3.9‰, and the percentage of the secondary market in Guangdong's total allowances trading volume increased remarkably to exceed 10%, though remaining at a low level. Without enough mobility, the carbon market will be unable to perform its normal investing and financial functions, which may somewhat offset the effect of carbon trading—an incentive marketization means—in saving energy and cutting emissions.

In order to increase carbon market breadth and mobility, Guangdong has been actively expanding the channels for both institutional and individual investors to assess to the market on the heels of the first institutional investor that entered Guangdong carbon market on March 14, 2014, the first batch of individual investors did so on 5 June.

Such practice as bringing in investors (nonregulated entities) is of great significance in increasing breadth and depth of carbon market. As of 2015-end, CGEX-registered institutional investors have traded 770,200 ton of allowances with total revenue of around 36.62 million yuan; while individual investors have traded 47,200 ton with total revenue of 1.98 million yuan. The combined trading volume and revenue of the two investors, respectively accounted for 29.09 and 26.25% of Guangdong's total trading volume and revenue. In short, Guangdong carbon market has been increasing varieties of participants and broadening market breadth.

8.4 Market Outlook and Suggestions

From the fixed-price allowances auction in 2013 to the present bidding mechanism, Guangdong has formed a primary carbon market for allowances auction with its own characteristics. Its secondary carbon market has drawn diverse participants. The overall carbon market of Guangdong has been expanding in both width and depth with the carbon pricing function deepening constantly.

At this point when the nationwide carbon market is about to be in full swing, Guangdong is actively introducing its distinctive ETS and attempting to link with the national market. In the meantime, there is a necessity for Guangdong to keep improving relevant policies and regulations, so as to provide better services to the low-carbon undertaking. We hereby put forward the following suggestions:

(1) Give priority to emissions reduction and stick to the core function of "carbon pricing"

Guangdong secondary carbon market has slack transactions and less active than other pilot carbon markets, which has limited profitability of speculators. The steady performance and positioning of Guangdong carbon market are closely associated with its functions in emissions cutbacks and price discovery. In our opinion, an important criterion for examining the validity of carbon market design is to see its effect in lowering the cost of cutting emissions and promoting low-carbon technologies. Any assessment that deviates from market positioning is worth of discussion. While making effort to realize the objectives of a carbon market, the interests of market participants are taken account and maximized, which is no doubt an optimal outcome. We believe that when low-carbon technologies in a certain area remain stable during a given period, the carbon market situation seems like a "Zero-sum Game", indicating that technical breakthroughs are required to improve environmental quality which will benefit the common population.

(2) Excessive expectation on functions of carbon market may inflict its healthy development.

Carbon market, based on clearly defined property, is in essence for achieving optimal allocation of scarce resources with an invisible hand. The definition of the total amount of allowances seems, especially, important, any perfunctory adjustment to the allowances that caters to the so-called "economic growth" is not allowed. In addition, some experts recommended to link PHCER with carbon market to enable regulated companies to purchase PHCER credits to fulfill their ETS obligations via a new path. But we believe that such opinion will increase unnecessary functions to carbon market, which will somewhat dampen the companies' activity in saving energy and cutting emissions. In this case, the carbon market may finally become an ornament, since large emitters remain unaffected, while the small emitters are pressed to cut emissions.

(3) Actively innovate carbon financial products with carbon pricing as basis.

The basic function of carbon market is to allocate emissions rights and interests with market-oriented means. Promoting financial resources to converge on the green and low-carbon sectors will continuously power the transition to a low-carbon society and benefit the in-depth development of a carbon market. Carbon financing is unable to develop unless there is a complete carbon market with the basic carbon pricing function, and capital input and withdrawal mechanisms. Guangdong has a bright prospect in this respect. Actually, the province has started innovating carbon financial products, e.g., allowances buyback, carbon asset management for corporate groups, allowances trusteeship, carbon business, carbon fund, conversion between allowances and CCER, and mid-to-long-term trading contract. We believe that there should be a green financing system that resembles IPO.

(4) Build a low-carbon development fund led by government but based on market-oriented operation.

It is advised that there shall be a provincial-level low-carbon development fund which is sponsored by government finance and incorporated with the current special funds for energy conservation and environmental protection, by following the principles of government guidance, industrial support, low-carbon impact, independent operation and risk control. We hereby present the following suggestions for such fund: (i) Diversify fund sources. The capital input at the preliminary stage shall come from the revenue from paid allocation of allowances. Expand channels (legitimate operating income and government subsidies) for drawing capital input. (ii) Most of the fund shall be diverted to support regulated companies to transform to low-carbon technologies, develop new energy and renewable energy-based projects and energy-saving projects. (iv) The funded projects are made up of "policy-initiated project" and "market-based project" which are subsidized with separate capital pool and subject to different decision-making mechanisms. Overall, the policy objectives of this carbon fund are to mobilize and consolidate carbon market resources, vitalize trading activities and expand market participation.

References

1. Weidong Liu. Guangdong allowances trading revenue exceeding 1 bln yuan, ranking first in many areas [EB/OL]. Nanfang Daily: http://economy.southcn.com/e/2016-05/13/content_147636109.htm.

2. Nanfang Daily. Allowances auction revenues to be injected into carbon fund for boosting emissions reduction [EB/OL]. http://epaper.southcn.com/nfdaily/html/2014-12/11/node_8.htm [2014-12-11].
3. Nanfang Weekly. Guangdong: allowances demand exceeding supply in the first auction of each year for four years in a row [EB/OL]. China Carbon Trading website: http://www.tanjiaoyi.com/article-18785-1.html.
4. Nanfang Daily. Guangdong carbon trading turnover ranked first in China, a nationwide carbon trading platform is brewing [EB/OL]. Clean Development Mechanism in China: http://cdm.ccchina.gov.cn/Detail.aspx?newsId=62452 [2016-7-14].
5. Jianjun Zhang, Guangdong carbon trading volume hitting 180 Mt, rising to the world third largest carbon market [EB/OL]. http://finance.sina.com.cn/china/dfjj/20150811/055022929891.shtml [2015-8-11].
6. Guangdong Provincial Development and Reform Commission. Guangdong Carbon Emission Allowance Management and Implementation Rules [EB/OL]. http://www.gd.gov.cn/govpub/bmguifan/201503/t20150324_210857.htm. [2015-2-16].

Chapter 9
Micro-impact Assessment of Guangdong ETS

9.1 Assessment of ETS Operational Efficiency

The ETS is an organic system that requires an all-round assessment from the perspective of systematology. In order to evaluate the administration efficiency of Guangdong ETS, our research team computed the relative efficiency of Guangdong ETS with DEA Model. The outcome is considered as a key criterion for judging appropriateness of Guangdong emissions allowances allocation. For employing this model, we used the 31 provinces/municipalities as the study samples, and 26 of them were defined Decision Making Unit (DMU) through screening.

In light of the *Notice of the State Council on Issuing the Work Plan for Greenhouse Gas Emissions Control during the 12th Five-Year Plan Period* released in 2012, there are 31 Chinese provinces/municipalities are about to cut carbon emissions, though to a different extent. After preliminary screening, we removed 6 of them (Inner Mongolia, Hainan, Tibet, Qinghai, Ningxia, and Xinjiang) from our study samples, since they do not need to cap the total amount of emissions, while the remaining 26 (Shenzhen Municipality and 25 provinces/ municipalities due to cut emissions) are defined as DMU for model-based evaluation. We used C^2GS^2 Model to evaluate the administration efficiency of the 26 areas [1, 2].

9.1.1 Model Building

Assumption:

Each ETS represents a unit, called as $DMU_i(1 \leq i \leq 26)$. Each DMU is constituted by six input factors (X) and four desirable output (Y)—Y is generated from ETS operation, then the Production Possibility Set (PPS) is defined as follows [3]:

© China Environment Publishing Group Co., Ltd. and Springer Nature Singapore Pte Ltd. 2019

D. Zhao et al., *A Brief Overview of China's ETS Pilots*,
https://doi.org/10.1007/978-981-13-1888-7_9

$$P = \{(X, Y), X > 0, Y > 0; X \in E_t, Y \in E_m\}$$

The input and output vector of the ith DMU is

$$X_i = (x_{1i}, x_{2i}, x_{3i}, \ldots, x_{ti})^{\mathrm{T}} > 0 \quad (1 < t \leq 6)$$
$$Y_{\mathrm{i}} = (y_{1i}, y_{2i}, y_{3i}, \ldots, y_{mi})^{\mathrm{T}} > 0 \quad (1 < m \leq 4) \ \ (1 \leq i \leq 26)$$

x_{ti} denotes the tth input indicator of the ith DMU; y_{mi} denotes the mth output indicator of the ith DMU. Both x_{ti} 和 y_{mi} are observed values calculated on the basis of historical statistics or predictions.

Based on the rationale of C^{2}GS^2 Model, bring non-Archimedean infinitesimal $\varepsilon(\varepsilon = 10^{-6})$ into the linear programming equation, then convert the equation into the Dual Model of LP:

$$\min \left\{\theta_k - \varepsilon \cdot \left(\widehat{e^T} S^- + e^T S^+\right)\right\}$$
$$\text{s.t.} \begin{cases} \sum_{i=1}^{26} \lambda_i \cdot X_i + S^- = \theta_k \cdot X_k \\ \sum_{i=1}^{26} \lambda_i \cdot Y_i - S^+ = Y_k, \\ \sum_{i=1}^{26} \lambda_i = 1, \lambda_i \geq 0 \\ S^+ \geq, S^- \geq 0 \end{cases} \tag{9.1}$$

where $\widehat{e^T} = (1, 1, \ldots, 1) \in E_n$, $e^T = (1, 1, \ldots, 1) \in E_m; S^-, S^+$, respectively, denote slack variable of input and output.

When $\theta = h(X_k, Y_k) = 1$, the input X_k can no longer decrease by the same ratio θ, then DMU has a technical efficiency (weak DEA efficiency); when $\theta = h(X_k, Y_k) = 1$ and $S^{k-} = S^{k+} = 0$, then DMU has DEA efficiency, indicating PPS is at the effective production frontier.

If $\theta \neq 1$, it indicates that technical efficiency of DMU is available for further improvement, then use a distance function to figure out the "projection" of DMU at the effective production frontier:

$$\widehat{X}_k = \theta X_k - S^-, \quad \widehat{Y}_k = \theta Y_k + S^+ \tag{9.2}$$

In order to obtain the adjusting information for converting it to technical efficiency. Through statistical analysis of adjustment vector, we will find out the major problems that affect the efficiency of DMU:

Adjustment vector of input: $\Delta X_k = X_k - \widehat{X}_k = (1 - \theta)X_k + S^-$
Adjustment vector of output: $\Delta Y_k = \widehat{Y}_k - Y_k - S^+$

9.1.2 Input–Output Indicator Design

Choosing appropriate indicators for assessment based on characteristics of study samples is one of the crucial steps for developing models. While considering the characteristics and construction objectives of ETS, we designed the input–output indicators used for C^2GS^2 calculation in this report. After a preliminary estimation and expert review, we sorted out the following six input indicators and four output indicators.

(1) **Input indicators**

We designed input indicators for the efficiency evaluation model based on the major components of an ETS. Through literature analysis, we have found eight indicators that are closely related to ETS operational efficiency: regulated emissions quantity, allowances total quantity and allocation, MRV, transaction system, offset mechanism, legal framework, and linkage with external carbon markets. We visited and consulted the stakeholders (ETS designers, researchers, government officials, and corporate representatives) in the pilot areas, and found that the current ETS pilot programs only involve the former six indicators, and they are at the beginning to link with each other. Among the six indicators, allowances allocation plan occupies the core position. Most pilot areas adopt the "top-down" and "bottom-up" approaches for determining the total amount of allowances, and the latter is closely related to allowances allocation. It is believed that the allocation plan somewhat affects allowances total quantity, carbon price, companies' efforts in emissions reduction and economic efficiency.

According to the on-site surveys of Guangdong and other ETS pilot areas, we learned that they have roughly the same frequency of allowances allocation. Guangdong was the only one that exercised mandatory allowances purchase in 2013 but turned to voluntary purchase in 2014. Moreover, through an initial model calculation, we found that the two factors (allocation frequency and mandatory/voluntary purchase) have no significant direct impact on evaluating the allocation efficiency with DEA Model. After removing these two factors, the model-based evaluation outcome turned out to be significant, so we used the following six indicators as input factors for model calculation.

X_1 percentage of the carbon emissions from ETS-regulated industrial sectors in local total emissions in year t_0 (2010);
X_2 comprehensive emissions reduction rate of ETS-regulated industrial sectors (criterion for allowances allocation);
X_3 percentage of the IVA of ETS-regulated industrial sectors in local total GDP;
X_4 allowances trading price;
X_5 percentage of CCER for offsets;
X_6 fines for excessive emissions.

The ETS framework in the pilot areas also covers newly built projects. The national industry development plan has presented requirements for the scale and technical norms of newly built projects, e.g., their CO_2 emissions per unit of production shall be lower than their existing counterparts, meaning that these projects only account for a small share in local total emissions. Therefore, all pilot areas usually allocate free and sufficient emission allowances to these projects, which exert limited impact on the ETS efficiency evaluation. So the newly built projects are excluded from the input factors for evaluation.

(2) **Output indicators**

Under the restraint of the same total amount of emissions, the ETS-regulated companies have lower emission reduction cost than those nonregulated companies, which is an essential benefit of the ETS; therefore, relative decline rate in emission reduction cost is a key output indicator for evaluating ETS. During the investigations, we also found that local governments are especially caring about the impact of ETS on local overall emission reduction target, GDP growth, and development of low-carbon service sector, which concerns the sustainable development of the ETS in China. Based on the above considerations, we took the following four output indicators for measuring the ETS efficiency.

(i) Relative decline rate in emission reduction cost (Y_1): The ETS features "lower cost in achieving emission reduction", thus making Y_1 an important factor for measuring the effect of the ETS in all pilot areas. Theoretically, Y_1 is calculated through comparison between the homogeneous companies within or outside the ETS framework.

$$Y_1 = \sum_{k=1}^{n} \frac{\overline{c_k}(a, e, p)}{\overline{C_k}(m, e, r)}$$

Under the prerequisite of control over carbon intensity and gross energy consumption, $\overline{C_k}$ 和 $\overline{c_k}$, respectively, denote the sectoral average emission reduction cost of sector K and the average emission reduction cost of ETS-regulated companies within sector K.

The parameters a, e, p, r, and m, respectively, stand for the amount of allowances, level of emissions, carbon price, potentials in emission reduction, and intensity of environmental constraint. However, when the ETS is yet officially operated, it is hard to compute Y_1 of ETS-regulated companies. As such, the above formula is better converted into the following formula:

$$Y_1' = \frac{\sum_k^n \overline{\mathrm{ac}_k}}{\overline{\mathrm{AC}}} = \frac{\sum_{k=1}^n (10.9 - 108 * \ln(1 - R))_k}{\overline{\mathrm{AC}}} \tag{9.3}$$

In Formula 9.3, $\overline{\mathrm{ac}_k}$ denotes the average emission reduction cost of the ETS-regulated sector K, $\overline{\mathrm{AC}}$ denotes the social average emission reduction cost, and R denotes the emission reduction rate of sector K. If the upper limit on emissions of ETS-regulated companies is above the social emission reduction target, then

$\overline{ac_k} > \overline{AC}$. In light of the ETS design rationale, the ETS-regulated companies are able to trade allowances to lower $\overline{ac_k}$ which may finally level off $\overline{AC}$. In this section, we use Formula 9.3 to expound the advantages of the ETS in cutting emission reduction cost.

(ii) ETS economic efficiency (Y_2): Measure Y_2 by comparing the growth rate of IVA of ETS-regulated companies and local GDP growth rate. $Y_2 > 1$ indicates ETS economic efficiency is higher than the sectoral average economic efficiency, which proves that the ETS is of small negative impact on the local economy.

$$Y_2 = \frac{\left(\sum_{k=1}^{n} \mathrm{GDP}_{k(t_1)} - \mathrm{GDP}_{k(t_0)}\right) / \sum_{k=1}^{n} \mathrm{GDP}_{k(t_0)}}{\left(\mathrm{GDP}_{(t_1)} - \mathrm{GDP}_{(t_0)}\right) / \mathrm{GDP}_{(t_0)}} \tag{9.4}$$

(iii) Developments of low-carbon sectors (Y_3): The design of ETS aims to foster new economic growth points, create more job opportunities and promote the development of low-carbon sectors. Y_3 is measured by a third-party verifier or carbon trading service agency on basis of the quantity (q), scale (s), and human sources (h) of new companies that are directly related to the ETS.

$$Y_3 = \frac{w_1 q_{i1} + w_2 s_{i2} + w_3 h_{i3}}{3} \quad i = 1, 2, \ldots, n \tag{9.5}$$

(iv) ETS contribution ratio to local emission reduction target (Y_4): Compare the decline rate in carbon intensity of the ETS-regulated sectors (Δe_k) and the local carbon intensity reduction target (Δe), we can figure out Y_4.

$$Y_4 = \frac{\Delta e_k}{\Delta e} = \frac{\sum_{k=1}^{n} \frac{E_{k(t_1)}}{\mathrm{GDP}_{k(t1)}} - \frac{E_{k(t_0)}}{\mathrm{GDP}_{k(t0)}}}{\frac{E_{t_1}}{\mathrm{GDP}_{t_1}} - \frac{E_{t_0}}{\mathrm{GDP}_{t_0}}} \tag{9.6}$$

9.1.3 *Input–Output Indicators*

In reference to the information and data in relevant statistical yearbooks and websites of competent authorities [4–12], as well as the industry materials and our survey findings, we used Formulas 9.3, 9.4, 9.5, and 9.6 to process and compute the data about 26 DMUs which are needed for model evaluation. Considering the limitations of coverage, we only introduced the input–output indicators in Table 9.1.

Table 9.1 Input–output indicators of ETS

Input indicators	Output indicators
Percentage of ETS-regulated emissions	Decline rate in emission reduction cost
Carbon intensity reduction target	Economic efficiency
Percentage of IVA of ETS-regulated companies	MRV employees
Percentage of CCER	Contribution ratio to emission reduction
Carbon price	
Fines for excessive emissions	

9.1.4 *Model Solving*

After being processed and computed, the input–output data are plugged into Formula 9.1, then we can apply EXCEL Programming Solver Function to resolve the respective operational efficiency of the 26 pilot areas:

Take Guangdong ETS for instance, the equation set for evaluating the impact of the allowance allocation plan on ETS operational efficiency is shown as follows:

$$\text{Objective function: } \min\left\{\theta_1 - \varepsilon\cdot\left(s_1^- + s_2^- + s_3^- + s_4^- + s_5^- + s_6^- + s_1^+ + s_2^+ + s_3^+ + s_4^+\right)\right\}$$

$$\text{Constraint condition}\begin{cases} 54\lambda_1 + 45\lambda_2 + 40\lambda_3 + 53.7\lambda_4 + + \cdots + 38\lambda_{26} + s_1^- = 54\theta_1 \\ 33\lambda_1 + 26.59\lambda_2 + 21\lambda_3 + 20\lambda_4 + + \cdots + 32\lambda_{26} + s_2^- = 33\theta_1 \\ 49\lambda_1 + 20\lambda_2 + 48\lambda_3 + 46.8\lambda_4 + \cdots + 59\lambda_{26} + s_3^- = 49\theta_1 \\ 44.3\lambda_1 + 59.2\lambda_2 + 20.3\lambda_3 + 59.3\lambda_4 + \cdots + 60.2\lambda_{26} + s_4^- = 44.3\theta_1 \\ 10\lambda_1 + 5\lambda_2 + 10\lambda_3 + 5\lambda_4 + \cdots + 10\lambda_{26} + s_5^- = 10\theta_1 \\ 132.8\lambda_1 + 50\lambda_2 + 60.9\lambda_3 + 59.2\lambda_4 + \cdots + 114\lambda_{26} + s_6^- = 132.8\theta_1 \\ 1.58\lambda_1 + 1.39\lambda_2 + 1.08\lambda_3 + 1.04\lambda_4 + \cdots + 1.53\lambda_{26} - s_1^+ = 1.58\theta_1 \\ \lambda_1 + 0.67\lambda_2 + 0.94\lambda_3 + 0.77\lambda_4 + \cdots + 1.07\lambda_{26} - s_2^+ = \theta_1 \\ 101\lambda_1 + 245\lambda_2 + 75\lambda_3 + 96\lambda_4 + \cdots + 318\lambda_{26} - s_3^+ = 101\theta_1 \\ 1.69\lambda_1 + 1.48\lambda_2 + 1.11\lambda_3 + 1.05\lambda_4 + \cdots + 1.64\lambda_{26} - s_4^+ = 1.69\theta_1 \end{cases} \quad (9.7)$$

Similar equations are applicable for evaluating the ETS operational efficiency of other pilot areas, but the right-hand side of the equations shall be replaced with the factors and θ_i of the concerned area.

9.1.5 *Outcome of Model Evaluation*

Tables 9.2 and 9.3 show that among the 26 DMUs, 15 have DEA efficiency, 2 have weak DEA efficiency, and 9 have non-DEA efficiency.

Among the seven ETS pilot areas, Beijing, Shenzhen, Chongqing, and Hubei have DEA efficiency, and the optimal solution of their ETS operational efficiency is given as follows:

Table 9.2 Calculation results of ETS operational efficiency and slack variables

Areas	θ	S_1^-	S_2^-	S_3^-	S_4^-	S_5^-	S_6^-	S_1^+	S_2^+	S_3^+	S_4^+
Guangdong	0.87	0.21	0.82	0.00	0.00	0.00	78.44	0.02	0.17	0.00	0.00
Beijing	1.00	0.00	0.00	0.00	0.00	0.00	0.00	0.00	0.00	0.00	0.01
Tianjin	0.88	0.00	0.25	5.69	0.00	3.80	0.00	0.00	0.00	2.83	0.01
Hubei	1.00	0.00	0.00	0.00	0.00	0.00	0.00	0.00	0.00	0.00	0.00
Shenzhen	1.00	0.00	0.00	0.00	0.00	0.00	0.00	0.00	0.00	0.00	0.00
Shanghai	0.90	17.33	0.00	0.03	0.00	0.00	0.00	0.00	0.05	0.00	0.04
Chongqing	1.00	0.00	0.00	0.00	0.00	0.00	0.00	0.00	0.00	0.00	0.00
Hebei	0.95	0.27	0.00	4.39	0.00	0.00	0.00	0.00	0.00	0.00	0.02
Shanxi	0.94	0.69	0.00	6.92	0.00	0.00	0.00	0.00	0.00	6.68	0.00
Liaoning	0.99	0.60	0.00	3.62	0.72	0.00	0.72	0.00	0.00	0.00	0.01
Heilongjiang	1.00	0.00	0.00	0.00	0.00	0.00	0.00	0.00	0.00	0.00	0.00
Jiangsu	1.00	0.00	0.00	0.00	0.00	0.00	0.00	0.00	0.00	0.00	0.00
Zhejiang	1.00	0.00	0.00	0.00	0.00	0.00	0.00	0.00	0.00	0.00	0.00
Anhui	1.00	0.00	0.00	0.00	0.00	0.00	0.00	0.00	0.00	0.00	0.00
Fujian	1.00	0.00	0.00	0.00	0.00	0.00	0.00	0.00	0.00	0.00	0.00
Jiangxi	0.93	0.00	0.00	0.00	0.25	0.00	0.21	0.01	0.00	0.00	0.00
Shandong	1.00	0.00	0.00	0.00	0.00	0.00	0.00	0.00	0.00	0.00	0.00
Henan	1.00	0.00	0.00	0.00	0.00	0.00	0.00	0.00	0.00	0.00	0.00
Hunan	1.00	0.45	0.00	0.00	0.00	0.00	0.00	0.00	0.00	0.00	0.00
Guangxi	1.00	0.00	0.00	0.00	0.00	0.00	0.00	0.00	0.00	0.00	0.00
Sichuan	1.00	0.00	0.00	0.00	0.00	0.00	0.00	0.00	0.00	0.00	0.00
Guizhou	1.00	0.00	0.00	0.00	0.00	0.00	0.00	0.00	0.00	0.00	0.00
Yunnan	1.00	0.05	0.00	0.00	0.02	0.00	0.02	0.00	0.00	0.22	0.00
Shanxi	0.97	0.00	0.17	5.97	0.00	0.12	0.00	0.00	0.00	1.42	0.01
Gansu	0.99	2.63	0.04	0.00	0.00	0.00	0.00	0.00	0.00	51.98	0.00

Table 9.3 Emission reduction cost of 300 MW generating units

Carbon price (yuan/tCO_2)	Percentage of paid allowances				Overall cost (yuan)
	3%	5%	10%	100%	
5	38	63	125	1251	1,094,210,000
60	451	751	1502	15,017	1,094,210,000
120	901	1502	3003	30,035	1,094,210,000

$\theta_1 = \theta_2 = \theta_{21} = \theta_{26} = 1$, $s_1^- = s_2^- = s_{21}^- = s_{26}^- = 0$, $s_1^+ = s_2^+ = s_{21}^+ = s_{26}^+ = 0$; therefore, DMU_2 (Beijing), DMU_4 (Hubei), DMU_5 (Shenzhen), and DMU_7 (Chongqing) have DEA efficiency (C^2GS^2), i.e., maximized output in contrast to input, and all of them lie at efficient production frontier.

As for the ETS in Guangdong, Shanghai, and Tianjin, they have input–output slack variable >0, indicating that the efficiency of both input and output shall be improved. Among the 19 DMUs, 11 have DEA efficiency, 6 have non-DEA efficiency, and 2 have weak DEA efficiency. The efficiency value of non-DEA efficiency is $\theta_1 < 1$, indicating that Hebei, Shanxi, Liaoning, Jiangxi, Shaanxi, and Gansu are now at the stage of progressively increasing scale. The ETS in Hunan and Yunnan have technical efficiency and weak DEA efficiency (Figs. 9.1 and 9.2).

Through analysis of the ETS operational efficiency of 26 DMUs, we found that Guangdong ETS had non-DEA efficiency over 2013–2014, which exhibits in the following three aspects:

(1) The CO_2 emissions covered by Guangdong ETS are overly high. In 2013, the regulated emissions accounted for 54% of Guangdong total emissions, and 12% points higher than the optimal efficiency point;
(2) In case of a high coverage of CO_2 emissions, the fines for excessive emissions will be on the high side, which further lowers allowance allocation efficiency. Through comparisons, we find that at the state of optimal efficiency, the fines for excessive emissions are around 37 yuan/t. In light of the carbon trading implementation plan of Guangdong, if an ETS-regulated company has excessive emissions, an amount of allowances that are twice more than the extra portion of emissions will be deducted from its next year's total allowances; in other words, it will be imposed a fine at 88 yuan/t which is two times more than carbon price and remarkably higher than efficiency value;
(3) Under the scenario that Guangdong ETS targets to reduce carbon intensity by 33%, covered enterprises will find the decline rate in emission reduction cost

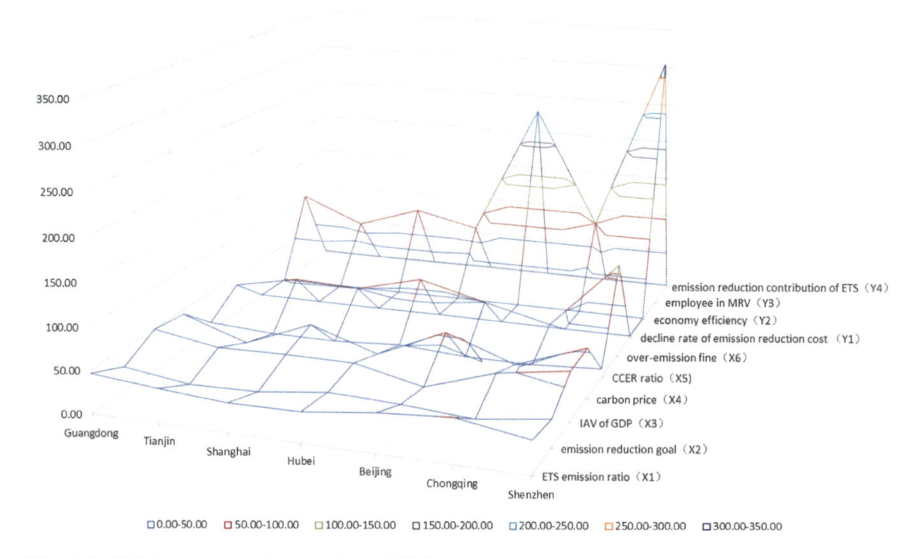

Fig. 9.1 Efficient production frontier of ETS operational efficiency of seven pilot areas

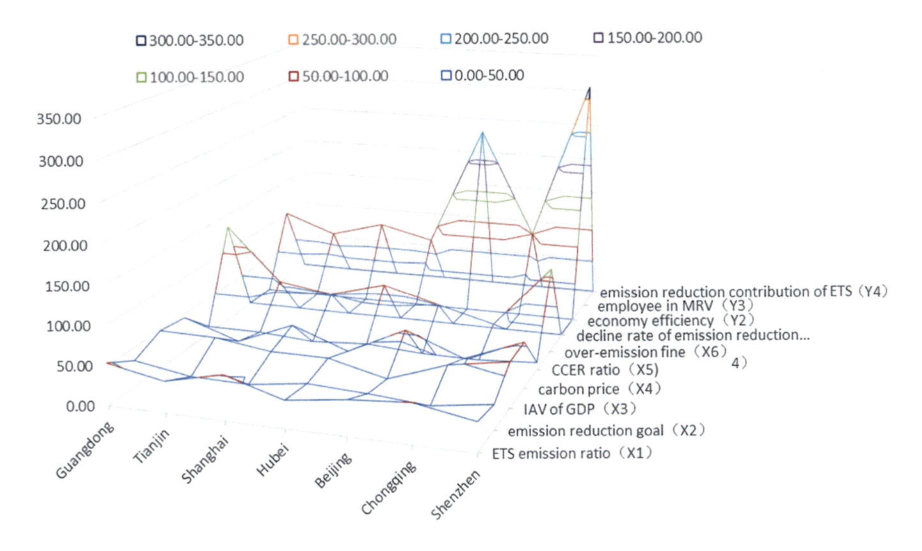

Fig. 9.2 Actual efficiency of seven pilots ETS

and economic efficiency deviate from efficiency value. The carbon intensity is about to be cut by 21.5%, then there will be higher economic efficiency (Fig. 9.3).

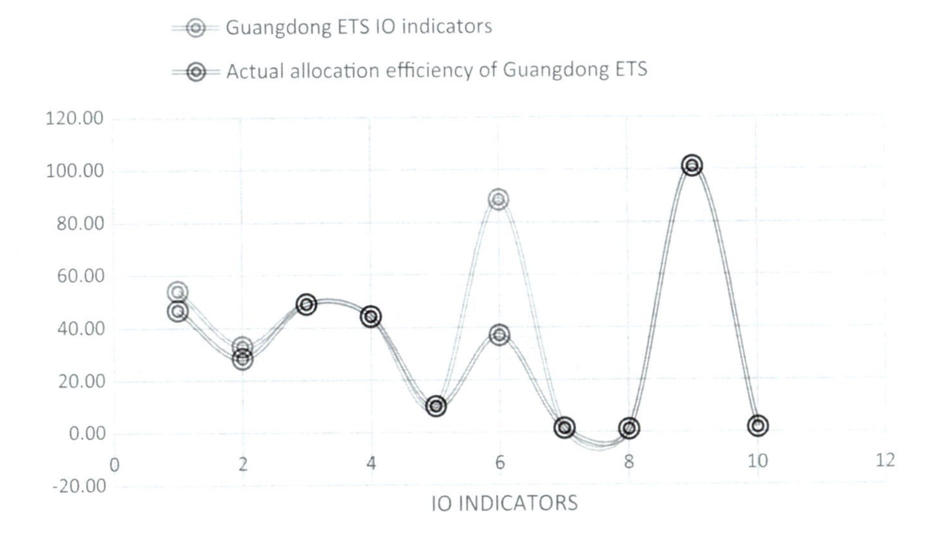

Fig. 9.3 Comparison between Guangdong ETS actual allocation efficiency and efficient production frontier

9.1.6 Analysis and Suggestions

Regarding the influencing factors of ETS operational efficiency, the quantity of regulated emissions, upper limit of emission allowances, covered enterprises' profitability and potentials in emission reduction play a crucial role in determining ETS operational efficiency. ETS operational efficiency is determined by its input–output scale and various factors, meaning that the operational efficiency of DMU is affected by various factors. For some DMUs, their regulation coverage is so wide that weighs on operational efficiency; for others, the covered enterprises are bearing overly high task for cutting emissions, which presses down the efficiency; or some loose allowance supply which is hard to restrict emissions.

Various input factors are closely correlated and interlocked, which means that an adjustment of the appropriateness of an input factor is related to other factors, there shall be no horizontal comparison of single factors or placing it under standardized processing. Take Guangdong and Chongqing, for instance, their ETS has similar input factors, and input–output value exceeding the average level, but Chongqing ETS has DEA efficiency, while Guangdong ETS has efficiency loss since it covers a large quantity of emissions. For the ETS of Guangdong and Chongqing, their regulated emissions account for more than 50% of local total emissions, and they have the same-level carbon prices and IVA of regulated companies, but as for the fines for over emissions, Guangdong ETS is twice more than Chongqing ETS, such a severe penalty deviates Guangdong ETS operational efficiency from its optimal scale. Therefore, all input factors of ETS are mutually confined and linked, so the ETS design shall closely focus on the characteristics of covered enterprises, and prepare for scale design of input factors (Fig. 9.4).

Percentage of CCER is closely related to allowance price, the percentage of available CCER is better adjusted down. Owing to sufficient supply and lower emission reduction cost, CCER price is usually lower than the allowance price. In a relatively closed market, if covered enterprises are able to buy a large quantity of CCER, they will have less demand for allowances, which will then bring down allowance price. Gloomy allowance price is sure to hold back the ETS from exhibiting its intrinsic advantage in saving emission reduction cost, because the companies that are highly potential in cutting emissions will lack motivation in investing emission reduction technologies or selling allowances amid low carbon prices, while the companies that are less potential in cutting emissions may purchase CCER for compliance in case of inadequate allowances, which will discount the ETS advantages and meaning of existence. Table 9.2 demonstrates that the ETS of Tianjin and Shanxi has a lower percentage of CCER projection value than its designed value at efficient production frontier. As mentioned above, a wholly applicable percentage of CCER does not exist, but it is feasible to adjust down (rather than instead of vice versa) such percentage.

The emissions covered by the ETS shall be no more than 50% of local total emissions at the preliminary stage. The emission reduction tasks assigned to the covered enterprises shall be properly designed in light of their characteristics,

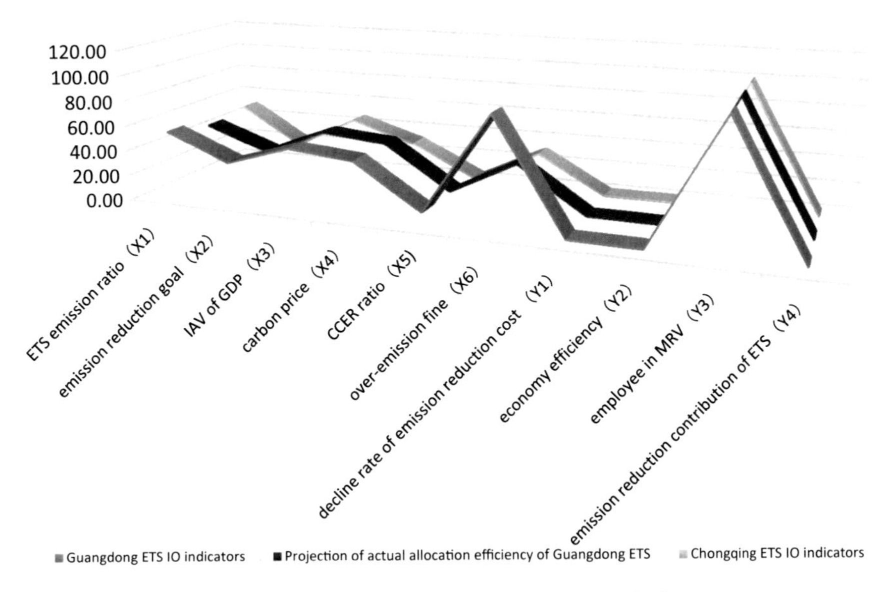

Fig. 9.4 Comparison between Guangdong and Chongqing ETS IO indicators

safeguard their normal development while urging them to save energy and cut emissions, so as not to dampen local economic growth.

Link allowance administration with CCER offset mechanism, i.e., in case of tight allowance supply, allow the covered enterprises are to turnover more CCER for offsetting their emissions; conversely, adjust down the percentage of available CCER. Set up an allowance reserve, motivate third-party institutions to join in the ETS so as to vitalize the carbon market.

In case the regulated industrial sectors function as an economic pillar of a given area, the upper limit on CO_2 emissions (emission reduction rate) should not be too tight, otherwise, it may discourage the sustainable development of local economy.

Any excessive emission shall be fined to demonstrate the deterrent effect of the ETS. In this sense, severe punishment helps to achieve the ETS objectives. However, when the ETS covers a large quantity of emissions, sets a tight upper limit on total emissions and has low percentage of available CCER, then the penalty mechanism for excessive emissions will not only be the sword of Damocles, but a sharp device for increased cost in emission compliance. An overly high amount of fine may inflict development of the real economy.

Fostering new economic growth, especially developing low-carbon service sector, is another important function of ETS. The operation of ETS calls for a large number of MRV institutions and professionals, which will create job opportunities and optimize local economic structure. Theoretically, more employees working for MRV prove that ETS is able to promote the development of the local low-carbon sector. But the employment scale of MRV differs among different areas with disparate labor cost, because the expenses on MRV are finally borne by the companies under verification. For the central and western regions where the per capita wage is relatively low, more personnel shall be trained to join in MRV; moreover, most of the employees in these regions lack of advanced knowledge or technical skills, cultivating more professional verifiers is able to enhance the accuracy of verification. For the eastern and other developed provinces, the scale of MRV should not be too large.

9.2 Impact of Allowance Allocation on Companies' Cost and Economic Efficiency

Under the condition of limited CO_2 emission space, the emission allowance allocation is actually a distribution of potential resources and wealth, which will exert an impact on macro-economy and companies' production cost from different dimensions. The design of the emission allowance allocation system shall take account of local emission reduction target, industry development demand, energy consumption mix, companies' technical strength and bearing capacity. Companies, being micro-economic agent, are the most directly under the impact from emission allowance allocation. Therefore, this section focuses on evaluating the impact of Guangdong ETS emission allowance allocation on companies' cost and economic efficiency [13].

9.2.1 *Comparative Analysis of Companies' Operational Cost*

The electricity and heat generation–supply sector (hereinafter briefed as "electricity sector") is a dominant GHG emitter and also a major ETS-regulated object. So all of the seven ETS pilot areas have included electricity sector into their regulation, and a study focused on this sector is widely representative. This sector, with Guangdong-based coal-fired electricity generating units as study samples, will evaluate the impact of the emission allowance allocation system on companies' economic efficiency, and sum up the crucial factors of the system that impact companies' operational cost.

There are four types of coal-fired generating units that are regulated by Guangdong ETS: <300, 300, 600, and 1000 MW. The CO_2 emissions from these four generating units were used as a benchmark for defining the allowances for Guangdong's total regulated coal-fired generating units in 2013. The 300 MW generating units have the largest number, the 600 MW generating units have the largest aggregated installed capacity; both the quantity and aggregated capacity of 1000 MW generating units are on the rise; the generating units less than 300 MW are being phased out. As such, the 300, 600, and 1000 MW generating units become study sample in this section. With a view to actual production, the power plants, for the sake of stable production and maximized efficiency, usually operate two generating units or more at the same time; and the production parameters and cost of the two generating units are assessed together. Therefore, in this sector, we make two generating units as one analytical unit for evaluating the ETS impact on the production cost of power plants.

(1) Operational cost breakdown of Guangdong-based regulated electricity companies

Our research team investigated the CO_2 emissions from Guangdong-based power plants, and consulted with the *Reference Construction Cost Index for Quota Design of Fossil Fuel-based Power Plant projects*, and found out that in 2014 the coal prices were about 800 yuan/tce, the limestone prices were about 100 yuan/t, the annual utilization hours of generating units were 4500 h, the electricity price was 0.502 yuan/kWh, the economic operational cycle of power plants were 20 year, the loan accounted for 80% of total static investment, the loan term averaged at 15 year, the loan interest rate averaged at 6.55%, the depreciable life was 15 year and the residuals rate was 5%, insurance and reparation expenses, respectively, held 0.25 and 2% of total investment, the employees' average wage was around 50,000 yuan/year (plus an additional 60% of welfare). The working efficiency of the sample generating units is an average efficiency of homogeneous generating units, the coal consumption per unit electricity output is converted from the benchmark value of allowances.

Power plants with 2 × 300 MW generating units: unit investment of 4394 yuan/kW; number of employees at about 234; material expense at about 6 yuan/kWh; other expenses at about 12 yuan/kWh; limestone consumption at 8 t/h (sulfur content of coal at about 2%); discharging fee of SO_2, NO_X and dust was, respectively, 1.43 million yuan/year, 1.62 million yuan/year, and 80,000 yuan/year. Based on our estimation, such power plants have an overall cost of about 1094.21 million yuan.

Power plants with 2 × 600 MW generating units: unit investment of 3367 yuan/kW; number of employees at about 247; material expense at about 5 yuan/kWh; other expenses at about 10 yuan/kWh; limestone consumption at 16 t/h (sulfur content of coal at about 2%); discharging fee of SO_2, NO_X and dust was, respectively, 2.60 million yuan/year, 2.93 million yuan/year and 150,000 yuan/year. Based on our estimation, such power plants have an overall cost of about 1913.10 million yuan.

Power plants with 2 × 1000 MW generating units: unit investment of 3334 yuan/kW; number of employees at about 300; material expense at about 4 yuan/kWh; other expenses at about 8 yuan/kWh; limestone consumption at 8 t/h (sulfur content of coal at about 0.9%); discharging fee of SO_2, NO_X and dust was, respectively, 3.60 million yuan/year, 4.10 million yuan/year and 240,000 yuan/year. Based on our estimation, such power plants have an overall cost of about 3032.55 million yuan.

The cost breakdown of the above three generating units is respective shown in Figs. 9.5, 9.6 and 9.7:

As shown in the above figures, the expenses on coal, depreciation, loan interest, reparation, and other items account for more than 95% of power plants' overall cost. Among all cost items, the expenses on coal rank first, which are about

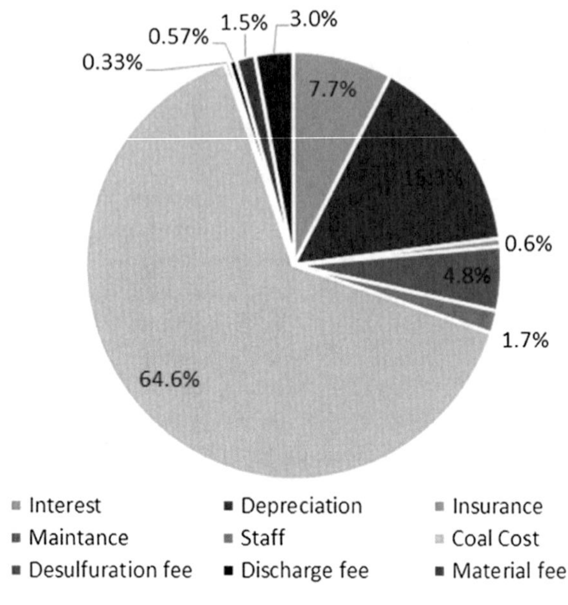

Fig. 9.5 Cost breakdown of 300 MW generating units

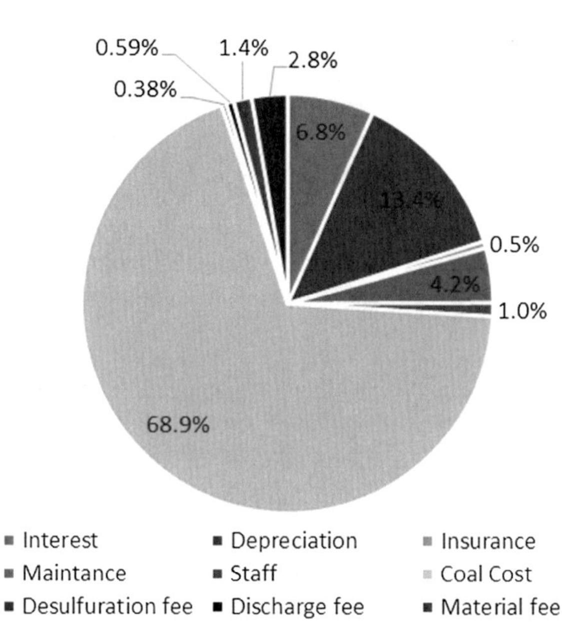

Fig. 9.6 Cost breakdown of 600 MW generating units

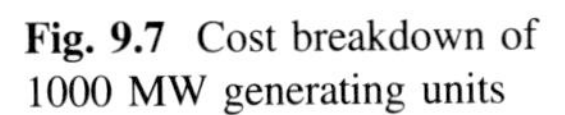

Fig. 9.7 Cost breakdown of 1000 MW generating units

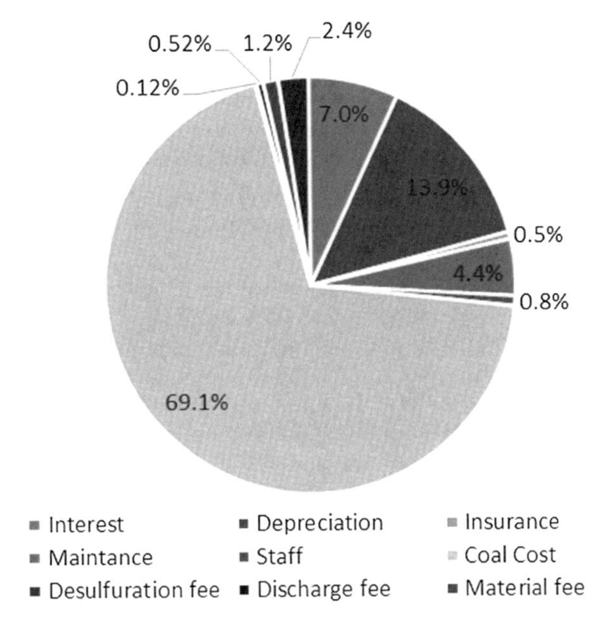

two-thirds of overall cost, indicating that coal consumption quantity and coal price are key factors that determine production cost of power plants. Along with expanding installed capacity, the marginal cost for labor force, material and administration has been decreasing progressively under the scale effect. Moreover, since the 1000 MW generating units require higher for feed coal quality (with much lower sulfur content), the percentage of desulfurization and discharge expenses are significantly lower than the other two generating units. However, coal consumption has kept increasing, holding an increasing share in the overall cost.

(2) Impact of allowance administration on the cost of Guangdong-based regulated electricity companies

Being an external environmental governance mechanism, ETS will bring a series of influences, both short term and long run, on covered enterprises. These companies may receive benefit from carbon trading, but pay for extra expenses in the meantime. In the short term, the companies have to increase manpower and material resources for making accounting of carbon emissions and taking care of carbon assets; purchase additional allowances if their actual emissions are much higher; or sell surplus allowances to gain profit in case of significant emission reductions, so as to offset all or partial of reduction cost. Different allowance allocation plans will incur different compliance costs upon covered enterprises. If 100% of the allowances are allocated for free, and the allowance benchmark is industrial average emissions, the ETS will exert a small impact on the cost of the power plants with production efficiency on industry average, such impact is almost ignored. While the emission administration is going deeper and carbon emissions are about to enter the peak time, the upper limit on aggregate emissions tends to be tightened, so as to ensure the fulfillment of the established reduction target. Meanwhile, the total amount of allowances shall be strictly managed. When the allowance benchmark remains unchanged, the percentage of free allowances shall be reduced; in this case, for the covered enterprises, if their quantity of emissions remains unchanged or the emission cutbacks are lower than the reduction in allowances, they have to purchase additional allowances from government or market, which will, of course, increase their cost.

In light of Guangdong carbon emission allowance allocation plans, and based on the study samples of selected power plants, we analyzed the impact of ETS on their production cost under different scenarios. We made the following assumptions:

(i) The emission reduction cost discussed in this section is limited to the direct expenses for the purchase of allowances, while disregard of such invisible cost for manpower and material resources injected into carbon trading;
(ii) Regarding the study samples, the energy consumption for electricity generation is an average level of the same generating units. When the allowance benchmark is based on the average energy consumption of the same generating units, the total allowances granted to the companies shall be equal to their actual emissions;
(iii) If the free allowances are lower than companies' actual emissions, they have to adopt certain emission reduction technologies to cut emissions or purchase additional allowances on market in order to fulfill their reduction obligation. But they have to pay for the expenses no matter what option they choose. There are diversified emission reduction technologies, which complicate the cost accounting, so we prefer estimating their reduction cost from the purchase of additional allowances;
(iv) The *Interim Measures for Carbon Emissions Administration in Guangdong Province* state that "the emission allowances granted to covered enterprises/institutions are based on free and paid allocation, with the percentage of free allocation reduced gradually," "electricity companies will receive 95% of free allowances, while iron and steel, petrochemical and cement companies will receive 97% of free allowances separately." In reference to the foreign experiences, e.g., EU ETS cancels free allowances to electricity sector (during Phase 2 and 3), and US RGGI never distributes allowances to electricity sector for free, so we estimate that Guangdong will gradually cut the percentage of free allowances, the electricity companies will have to buy 3, 5, 10, and 100% of allowances step by step;
(v) In 2013, Guangdong set the carbon price at 60 yuan/tCO_2 for auction. The carbon prices vary greatly among the seven pilot areas, some top at 120 yuan/tCO_2 (China Shenzhen Emission Exchange), while some lower at 5 yuan/tCO_2 (Shanghai Environment and Energy Exchange). As for Guangdong-based electricity companies, the carbon prices for different scenarios are set as 5, 60, and 120 yuan/tCO_2.

In light of the above assumptions and the characteristics of electricity sector, we developed a formula for calculating the emission reduction cost:

$$\begin{aligned}\text{Emission-reduction cost} &= \text{installed capacity} \times \text{generating hours} \\ &\quad \times \text{allocation benchmark} \times \text{percentage of paid allowances} \times \text{carbon price}\end{aligned}$$

The computed emission-reduction costs of the above-mentioned three generating units are shown in Tables 9.4 and 9.5 and Figs. 9.8, 9.9, and 9.10.

Table 9.4 Emission reduction cost of 600 MW generating units

Carbon price (yuan/tCO_2)	Percentage of paid allowances				Overall cost (yuan)
	3%	5%	10%	100%	
5	70	117	234	2336	1,913,100,000
60	841	1401	2803	28,026	1,913,100,000
120	1682	2803	5605	56,052	1,913,100,000

Table 9.5 Emission reduction cost of 1000 MW generating units

Carbon price (yuan/tCO_2)	Percentage of paid allowances				Overall cost (yuan)
	3%	5%	10%	100%	
5	111	186	371	3713	3,032,550,000
60	1337	2228	4455	44,550	3,032,550,000
120	2673	4455	8910	89,100	3,032,550,000

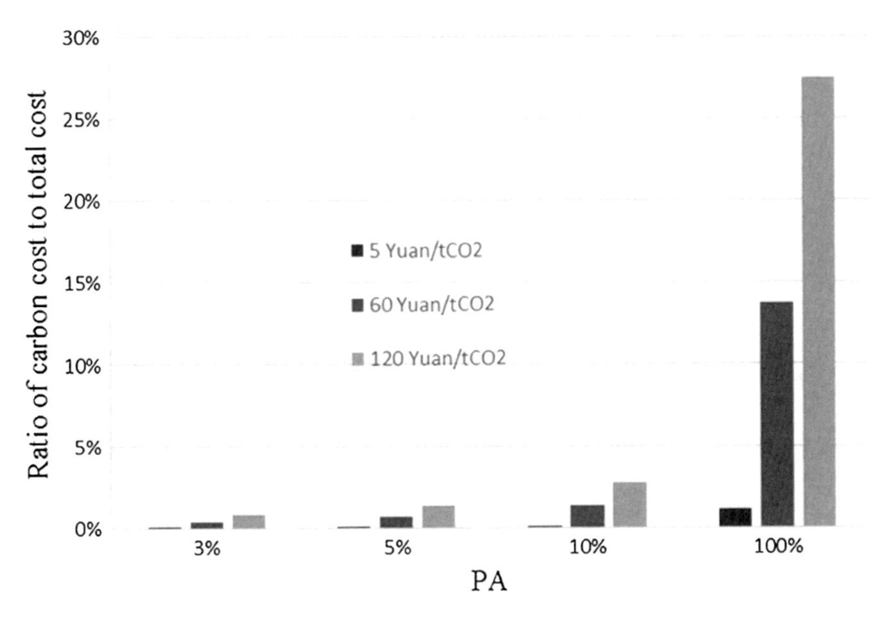

Fig. 9.8 Share of emission reduction cost in overall cost of 300 MW generating units

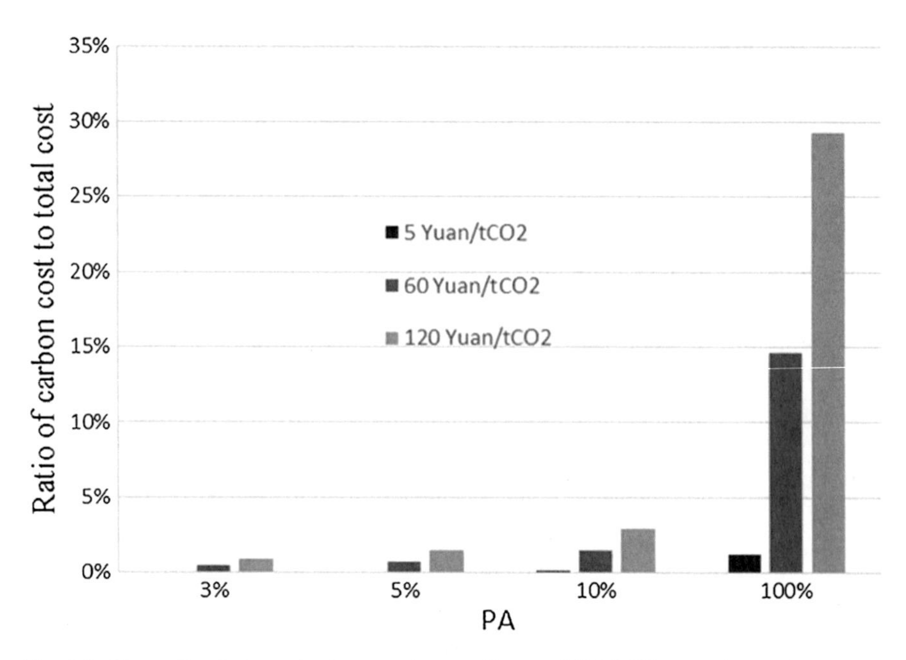

Fig. 9.9 Share of emission reduction cost in overall cost of 600 MW generating units

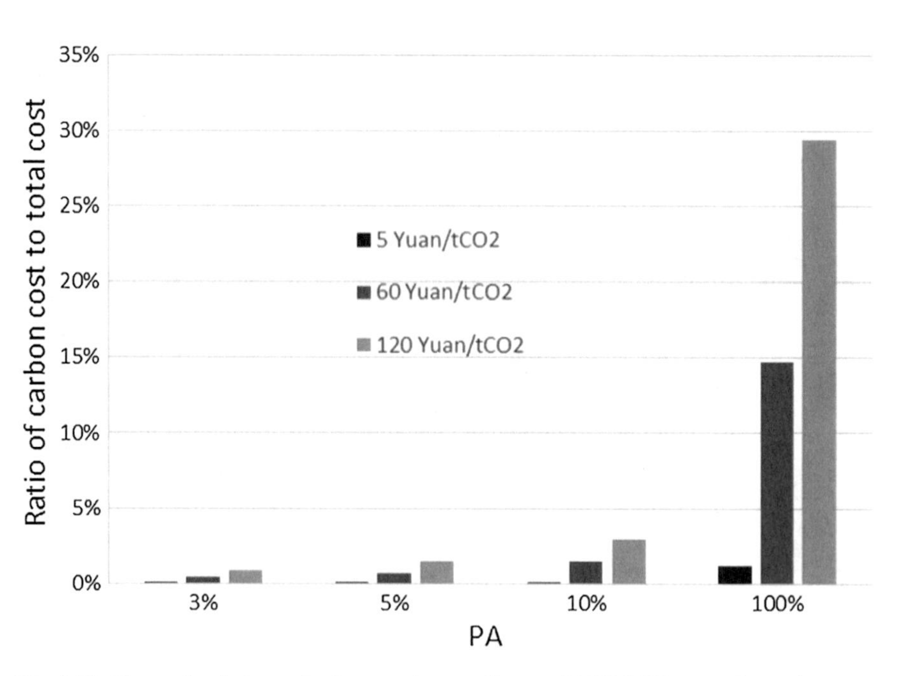

Fig. 9.10 Share of emission reduction cost in overall cost of 1000 MW generating units

We can draw up the following conclusions from the above emission reduction costs:

In case of low carbon price (5 yuan/tCO_2), the emission reduction cost accounts for a small share in overall electricity generation cost. Even if the company has to buy 100% of allowances, the emission reduction cost only holds around 1% of the overall cost, indicating that the ETS is of limited impact on the overall cost of electricity companies.

In case of high carbon price (120 yuan/tCO_2), the emission reduction cost accounts for a larger share in overall electricity generation cost. If the company has to pay for 3% of allowances, the emission reduction cost only holds around 1% of the overall cost; if the company buys 100% of allowances, then the share of the emission reduction cost will rise to one-third, becoming the second largest cost item next to coal consumption, indicating that the ETS is of significant impact on the overall cost of electricity companies. It can be seen that the impact of ETS on the cost of companies is closely related to the carbon price and amount of free allowances. When the level of emissions is fixed, less free allowances will force the companies to buy more allowances; at this point, the expenses on allowances are decided by carbon price: low carbon price saves the emission reduction cost and exerts limited impact on companies' overall cost; on the contrary, high carbon price greatly affects companies' overall cost. In the case of high carbon prices and the company needs to buy a large quantity of allowances, its production cost will increase sharply, which may even alter its production decisions. In light of the Guangdong allowance allocation plan, the electricity companies only access to a certain portion of free allowances, they have to pay for an additional 3–5% allowances. The carbon price in 2013 and 2014 start was set at around 60 yuan/tCO_2. Based on such price and the demand for allowances, for Guangdong-based regulated electricity companies, their emission reduction cost accounts for around 0.5% of their overall cost, such proportion levels off that of desulfurization and discharge expenses, which is of limited impact on their overall cost.

Since electricity sector is the largest emitting source, their free allowances are less than the other three sectors. The ETS-based cap on total allowances may have the most significant impact on the electricity sector. Based on the above analysis, under Guangdong ETS framework, electricity companies have to pay for certain emission reduction cost, which is of limited impact on their overall cost. Such conclusion accords with our investigations of Guangdong-based electricity companies.

9.2.2 *Impact on IRR of Generating Units*

Subsection 9.2.1 analyzes the static impact of emission reduction cost on companies' overall cost. For understanding the dynamic impact of emission reduction cost on the entire production cycle, we hereby use Internal Rate of Return (IRR) to compare and analyze the changes on IRR of the power plants with three types of generating units (2 × 300 MW, 2 × 600 MW, 2 × 1000 MW) based on the assumptions (i), (ii), and (iii) in Subsect. 9.2.1, in an aim to find the acceptable emission reduction cost which sustains normal profitability of different power plants. In order to gain an insight into the ETS impact on companies' economic benefit under different upper limits on carbon emissions, we split the percentage of paid allowances into seven grades (3, 5, 10, 20, 30, 50, and 100%) and divide the carbon prices into 36 multistep prices from 5 to 300 yuan/tCO_2, i.e., a price range split by 5 yuan/tCO_2 below 60 yuan/tCO_2 and a price range split by 10 yuan/tCO_2 above 60 yuan/tCO_2.

Based on above-mentioned parameters of generating units, we estimated the cash inflow and outflow of the electricity companies within their operating cycle, and then used EXCEl tools to figure out the IRR of different power plants (2 × 300 MW, 2 × 600 MW, 2 × 1000 MW) without ETS, i.e., 8.72, 15.58, and 17.65%. According to the *Reference Construction Cost Index for Quota Design of Fossil Fuel-based Power Plant projects*, the Baseline IRR of fossil fuel-based power plants is 8%, which is to say that if IRR > 8%, the power plant has economic viability. This shows that, without emission reduction cost, these power plants have an IRR > 8%, which is worthy of investment. In contrast, the implementation of ETS incurs emission reduction cost upon the power plants and affects their IRR. Based on different carbon prices and percentage of paid allowances, we compared calculated and analyzed the changes in IRR of these power plants.

In light of the assumed percentage of paid allowances (PA) and carbon prices, we can figure out the emission reduction cost of a power plant under a given situation. Based on the increase in emission reduction cost and changes on cash flow, we can compute the IRR of this power plant under such scenario. See the calculation results in Tables 9.6, 9.7, and 9.8.

Tables 9.6, 9.7, and 9.8 demonstrate that for the power plants with 300 MW generating units, in case of PA = 3%, carbon price >180 yuan/tCO_2; PA = 5%, carbon price >110 yuan/tCO_2; or PA = 10%, carbon price >55 yuan/tCO_2, their IRR is lower than Baseline IRR. In another case, for the power plants (600 and 1000 MW), based on PA = 3%, PA = 5% or PA = 10%, even if the carbon price is as high as 300 yuan/tCO_2, their IRR is much higher than the Baseline IRR.

In case of PA = 20%, carbon price >262 yuan/tCO_2, the 600 MW power plants have an IRR lower than the Baseline IRR, but the 1000 MW power plants still have economic viability. In case of PA = 30%, carbon price >236 yuan/tCO_2, the 1000 MW power plants will have an IRR lower than the Baseline IRR.

In case of PA = 100%, the 300 MW power plants (carbon price > 5.5 yuan/tCO_2), 600 MW power plants (carbon price > 52 yuan/tCO_2) and 1000 MW power plants (carbon price > 71 yuan/tCO_2) will have an IRR lower than the Baseline IRR and lost economic viability.

Through illustration of the changes on IRR based on different PA–carbon price portfolios, we have funded the relationship between carbon price, the percentage of paid allowances, and IRR, which reflect the ETS impact on companies' IRR.

Figures 9.11, 9.12, and 9.13 demonstrate that companies' IRR declines along with increasing emission reduction cost.

Table 9.6 IRR of 300 MW generating units with varied PA and carbon price portfolios

Carbon price (yuan/ tCO_2)	IRR						
	PA = 3%	PA = 5%	PA = 10%	PA = 20%	PA = 30%	PA = 50%	PA = 100%
5	*8.7%*	*8.7%*	*8.7%*	*8.6%*	*8.5%*	*8.4%*	*8.1%*
10	8.7%	8.7%	8.6%	8.5%	8.3%	8.1%	7.4%
15	8.7%	8.6%	8.5%	8.3%	8.1%	7.7%	6.7%
20	8.6%	8.6%	8.5%	8.2%	7.9%	7.4%	6.0%
25	8.6%	8.6%	8.4%	8.1%	7.7%	7.1%	5.3%
30	8.6%	8.5%	8.3%	7.9%	7.5%	6.7%	4.5%
35	8.6%	8.5%	8.3%	7.8%	7.3%	6.4%	3.7%
40	8.6%	8.5%	8.2%	7.7%	7.1%	6.0%	2.9%
45	8.5%	8.4%	8.1%	7.5%	6.9%	5.6%	2.1%
50	8.5%	8.4%	8.1%	7.4%	6.7%	5.3%	1.2%
55	8.5%	8.4%	8.0%	7.3%	6.5%	4.9%	0.2%
60	8.5%	8.3%	7.9%	7.1%	6.3%	4.5%	-0.8%

Note The figures (italic) represent the PA–carbon price portfolio under the scenario IRR > 8%

Table 9.7 IRR of 600 MW generating units with varied PA and carbon price portfolios

Carbon price (yuan/ tCO_2)	IRR						
	PA = 3%	PA = 5%	PA = 10%	PA = 20%	PA = 30%	PA = 50%	PA = 100%
5	*15.6%*	*15.5%*	*15.5%*	*15.4%*	*15.4%*	*15.2%*	*14.9%*
10	*15.5%*	*15.5%*	*15.4%*	*15.3%*	*15.2%*	*14.9%*	*14.2%*
15	*15.5%*	*15.5%*	*15.4%*	*15.2%*	*15.0%*	*14.6%*	*13.5%*
20	*15.5%*	*15.4%*	*15.3%*	*15.0%*	*14.8%*	*14.2%*	*12.8%*
25	*15.5%*	*15.4%*	*15.2%*	*14.9%*	*14.6%*	*13.9%*	*12.1%*
30	*15.5%*	*15.4%*	*15.2%*	*14.8%*	*14.4%*	*13.5%*	*11.4%*
35	*15.4%*	*15.3%*	*15.1%*	*14.6%*	*14.2%*	*13.2%*	*10.7%*
40	*15.4%*	*15.3%*	*15.0%*	*14.5%*	*14.0%*	*12.8%*	9.9%
45	*15.4%*	*15.3%*	*15.0%*	*14.4%*	*13.7%*	*12.5%*	9.2%
50	*15.4%*	*15.2%*	*14.9%*	*14.2%*	*13.5%*	*12.1%*	*8.1%*
55	15.4%	15.2%	14.8%	14.1%	13.3%	11.8%	7.6%
60	15.3%	15.2%	14.8%	14.0%	13.1%	11.4%	6.7%
70	15.3%	15.1%	14.6%	13.7%	12.7%	10.7%	5.0%
80	15.3%	15.0%	14.5%	13.4%	12.3%	9.9%	3.1%
90	15.2%	15.0%	14.4%	13.1%	11.8%	9.2%	0.9%
100	15.2%	14.9%	14.2%	12.8%	11.4%	8.0%	-1.6%

Note The figures (italic) represent the PA–carbon price portfolio under the scenario IRR > 8%

Table 9.8 IRR of 1000 MW generating units with varied PA and carbon price portfolios

Carbon price (yuan/ tCO_2)	IRR						
	PA = 3%	PA = 5%	PA = 10%	PA = 20%	PA = 30%	PA = 50%	PA = 100%
5	*17.6%*	*17.6%*	*17.6%*	*17.5%*	*17.5%*	*17.3%*	*17.0%*
10	*17.6%*	*17.6%*	*17.5%*	*17.4%*	*17.3%*	*17.0%*	*16.4%*
15	*17.6%*	*17.6%*	*17.5%*	*17.3%*	*17.1%*	*16.7%*	*15.8%*
20	*17.6%*	*17.5%*	*17.4%*	*17.1%*	*16.9%*	*16.4%*	*15.1%*
25	*17.6%*	*17.5%*	*17.3%*	*17.0%*	*16.7%*	*16.1%*	*14.5%*
30	*17.5%*	*17.5%*	*17.3%*	*16.9%*	*16.5%*	*15.8%*	*13.8%*
35	*17.5%*	*17.4%*	*17.2%*	*16.8%*	*16.3%*	*15.4%*	*13.1%*
40	*17.5%*	*17.4%*	*17.1%*	*16.6%*	*16.1%*	*15.1%*	*12.5%*
45	*17.5%*	*17.4%*	*17.1%*	*16.5%*	*15.9%*	*14.8%*	*11.8%*
50	*17.5%*	*17.3%*	*17.0%*	*16.4%*	*15.8%*	*14.5%*	*11.1%*
55	*17.4%*	*17.3%*	*17.0%*	*16.3%*	*15.6%*	*14.1%*	*10.4%*
60	*17.4%*	*17.3%*	*16.9%*	*16.1%*	*15.4%*	*13.8%*	*9.6%*
70	*17.4%*	*17.2%*	*16.8%*	*15.9%*	*15.0%*	*13.1%*	*8.0%*
80	17.3%	17.1%	16.6%	15.6%	14.6%	12.5%	6.5%
90	17.3%	17.1%	16.5%	15.4%	14.2%	11.8%	4.8%
100	17.3%	17.0%	16.4%	15.1%	13.8%	11.1%	3.0%
110	17.2%	17.0%	16.3%	14.8%	13.4%	10.4%	0.9%
120	17.2%	16.9%	16.1%	14.6%	13.0%	9.6%	-1.6%

Note The figures (italic) represent the PA–carbon price portfolio under the scenario IRR > 8%

If PA = 3%, the IRR curve tilts slightly when carbon price <150 yuan/tCO_2, it still look likes a straight line, it does not tilt downwards until carbon price >150 yuan/ tCO_2.
If PA = 5%, the IRR curve tilts slightly when carbon price <100 yuan/tCO_2, it does not tilt downwards until carbon price >100 yuan/tCO_2.
If PA > 10%, the slope of IRR curve changes notably along with rising carbon price.

The impact of ETS on IRR is related to the installed capacity of generating units. Our calculations show that the larger the installed capacity is, the smaller the IRR slope will be; a higher initial IRR (IRR without implementation of ETS) marks the company is able to bear higher emission reduction cost.

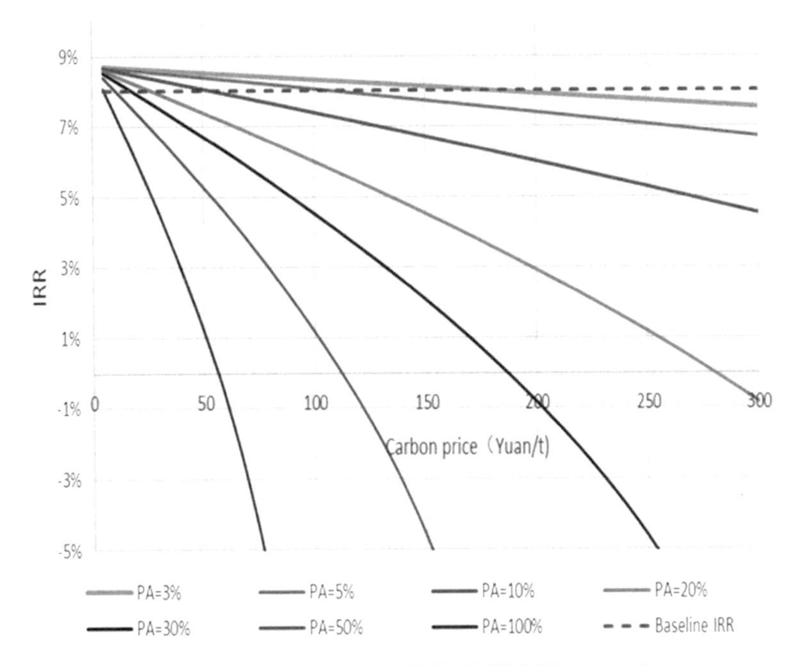

Fig. 9.11 Impact of emission reduction cost on IRR of 300 MW power plants

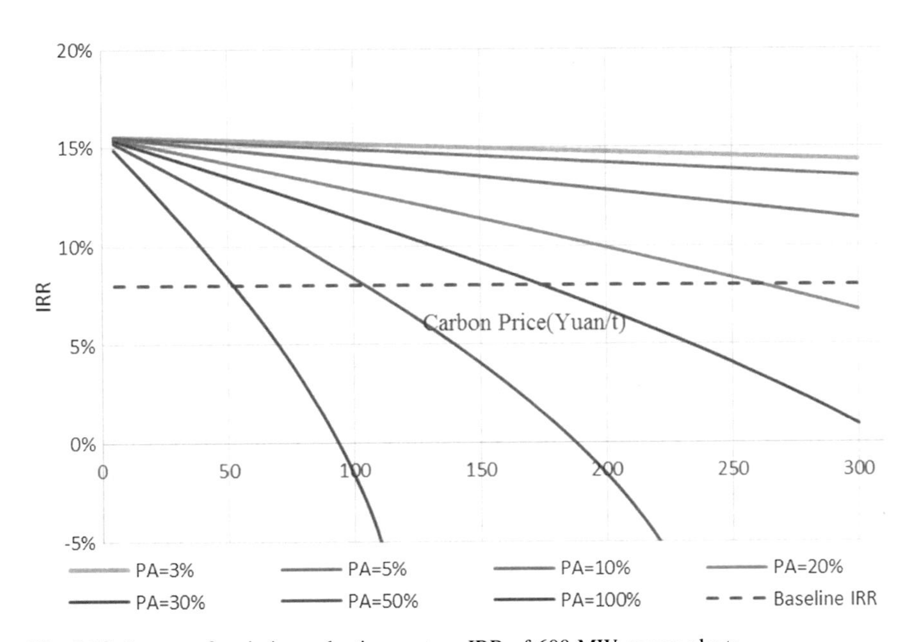

Fig. 9.12 Impact of emission reduction cost on IRR of 600 MW power plants

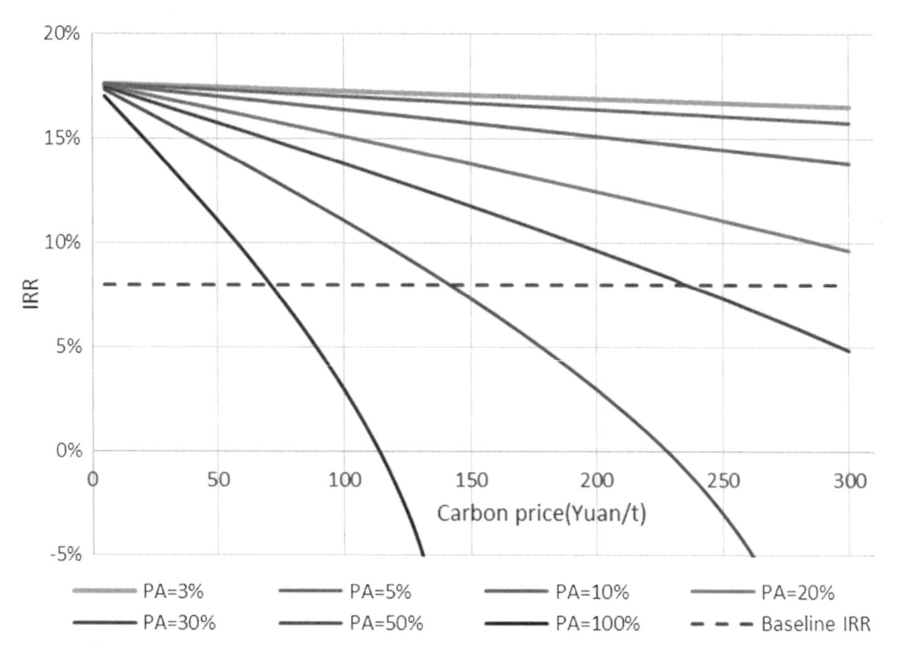

Fig. 9.13 Impact of emission reduction cost on IRR of 1000 MW power plants

9.2.3 *Critical Threshold of Electricity Companies for Bearing Emission-Reduction Cost*

In reference to the above analysis about the power plants (300, 600, and 1000 MW), given the paid allowances (PA) and carbon prices, the IRR of the three types of power plants can operate economically (IRR = 8%) and drew up Fig. 9.14 to exhibit the critical point where the emission reduction cost could affect companies' profitability.

For the 300 MW power plants, the PA–carbon price portfolios are set as (3%, 180 yuan/tCO_2), (5%, 110 yuan/tCO_2), (10%, 55 yuan/tCO_2), (20%, 27 yuan/tCO_2), (30%, 18 yuan/tCO_2), (50%, 11 yuan/tCO_2), and (100%, 5.5 yuan/tCO_2), and IRR as 8%. The dots on the curve indicate the maximized acceptable PA for maintaining IRR > 8% at a carbon price. The dots on the bottom left curve indicate the PA–carbon price portfolios, where the IRR > 8%; the dots on upper right curve indicate the PA–carbon price portfolios where the IRR < 8%.

By analogy, for the 600 MW power plants, the PA–carbon price portfolios are set as (20%, 262 yuan/tCO_2), (30%, 175 yuan/tCO_2), and (50%, 105 yuan/tCO_2), and IRR as 8%. For the 1000 MW power plants, the PA–carbon price portfolios are set as (30%, 236 yuan/tCO_2), (50%, 142 yuan/tCO_2), and (100%, 71 yuan/tCO_2), and IRR as 8%. Based on these dots, we are able to draw curves that depict the impact of emission-reduction cost on companies' economic efficiency (see Fig. 9.14).

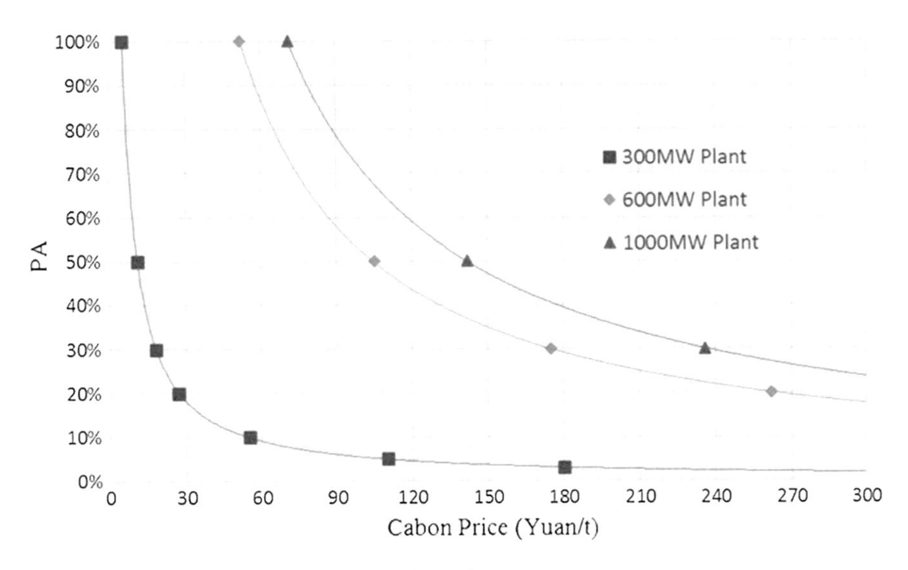

Fig. 9.14 Bearable PA of different generating units

We can use the above curves to compare the impact of emission-reduction cost on different power plants with varied PA–carbon price portfolios. In order to sustain normal profitability in case of carbon price = 90 yuan/tCO_2, the 300 MW power plants can buy no more than 7% of allowances, such percentage for the 600 and 1000 MW power plants is 58 and 79%, respectively. If a power plant has to buy 30% of allowances and maintains profitability in the meantime, the 300 MW power plants can bear a carbon price at no higher than 20 yuan/tCO_2, such price for the 600 and 1000 MW power plants is 175 and 238 yuan/tCO_2, respectively.

Moreover, Fig. 9.14 also demonstrates that the power plants with smaller installed capacity and higher unit energy consumption are harder to maintain IRR > 8%, and weaker to bear high emission reduction cost. From 300 to 1000 MW, the generating units have increasing generation efficiency, and a critical curve farther away from the axis, a wider range of profitability, and stronger to bear high emission reduction cost. If carbon price ≤ 60 yuan/tCO_2, the PA shall be lower than 10% to ensure economic viability of all generating units. Along with increasing carbon price, PA shall decrease accordingly.

The allowance allocation adopted by US RGGI is similar to the PA = 100% scenario discussed in this subsection. From 2010 to 2015, the average auction price set by RGGI was $1.86, $1.89, $1.93, $2.92, $4.72, and $6.10. Through comparison, we found that the RGGI auction prices at the preliminary stage are similar to the bearable carbon price (5.5 yuan/tCO_2) of Guangdong-based 300 MW power plants. From 2013 to 2015, the RGGI auction prices are similar to the bearable carbon price (52 yuan/tCO_2) of Guangdong-based 600 MW power plants. An evaluation report shows that the RGGI-regulated emissions over 2011–2013 dropped 19.8% from the 2006–2008 period, with unit emissions decreasing

141 $kgCO_2/MWh$. Such fact tells that within the RGGI framework, the unit emissions shall decrease along with an annual reduction in total emissions. As for Guangdong-based power plants (300, 600 MW), their unit emissions also demonstrate a year-on-year downward trend, which proves that our analysis of the bearable carbon price of power plants is reasonable on the whole.

References

1. LIU Yong, LI Zhixiang, LI Jing. Comparative study on DEA methods of environmental efficiency measurement [J]. Mathematics in practice and theory, 2010, 40(1):84–92.
2. WEI Quanlin, YUE Ming. DEA model introduction and analysis I. [J]. Systems Engineering-theory & Practice. 1998, (1):58–69.
3. WANG Wenjun, FU Chonghui, LUO Yuejun, XIE Pengcheng, ZHAO Daiqing. Management efficiency assessment of seven Chinese pilots carbon trading systems [J]. China Environmental Science, 2014, 34(6):1614–1621.
4. Beijing Municipal Commission of Development and Reform. The Notice on Carrying out the Work for the Carbon Emissions Trading Pilot Program in Beijing Municipality [EB/OL] http://www.bjpc.gov.cn/tztg/201311/t7020680.htm, 2013-11-22.
5. People's Government of Guangdong Province. The Interim Measures for Carbon Emissions Administration in Guangdong Province [EB/OL]. http://zwgk.gd.gov.cn/006939748/201401/t20140117_462131.html, 2014-1-15.
6. People's Government of Guangdong Province. The Plan for Carrying out the Work about the Carbon Emissions Trading Pilot Program in Guangdong Province [EB/OL]. http://zwgk.gd.gov.cn/006939748/201209/t20120914_343489.html, 2012-9-7.
7. People's Government of Hubei Province. The Plan for Carrying out the Work about the Carbon Emissions Trading Pilot Program in Hubei Province [EB/OL]. http://www.hbepb.gov.cn/zwgk/zcwj/szfwj/201303/t20130304_59436.html, 2013-2-18.
8. Shanghai Municipal People's Government. The Interim Measures for Carbon Emissions Administration in Shanghai Municipality (No. SMPC [2013] No. 10) [EB/OL]. http://www.shanghai.gov.cn/shanghai/node2314/node2319/node2404/n31728/n31729/u26ai37498.html, 2013-11-18.
9. General Office of Tianjin Municipal People's Government. The Plan for Carrying out the Work about the Carbon Emissions Trading Pilot Program in Tianjin Municipality [EB/OL]. http://www.tj.gov.cn/zwgk/wjgz/szfbgtwj/201303/t20130304_188946.htm, 2013-2-5.
10. General Office of Tianjin Municipal People's Government. The Interim Measures for the Administration of Carbon Emissions Trading in Tianjin Municipality [EB/OL]. http://www.tj.gov.cn/zwgk/wjgz/szfbgtwj/201312/t20131224_227448.htm, 2013-12-24.
11. Meiying Li, Beijing launching ETS pilot program, proposing mandatory regulation of 600 emitters [EB/OL]. http://finance.sina.com.cn/roll/20120403/025611740880.shtml, 2012-4-3.
12. Chafu Wulan. The Report of Low-carbon Development in Shenzhen (Shenzhen Green Book 2013) [M]. Shenzhen: Haitian Press, 2013.
13. Stephen Roosa, International Solutions to Sustainable Energy[M], Policies & Applications, The Fairmont Press, Inc and Taylor & Francis Group LLC, 2017.

Chapter 10
Impact of Guangdong ETS on Macroeconomy

10.1 Macro-evaluation of Carbon Trading Policies

Policy evaluation is an analysis and study of the policy scheme's effect and efficiency using some methodologies during formulation and implementation of regional policies or afterward, which is for the purpose of improving rationality and effect of the policy. By setting up a series of evaluation criteria and indicators, policy evaluation is able to inform policy-makers and the public of the implementation effect of emissions reduction policies, and demonstrate the essence of such policies and commitments of the proponents.

Being a market-oriented environmental policy instrument, the ETS is already executed in several countries and regions. In order to make a scientific and objective evaluation of the ETS effect, both domestic and foreign researchers, from different perspectives, have been evaluating the implementation of the carbon trading policies.

There is a sequential evaluation of the policies, such as from the beforehand, halfway, and afterward. The evaluations, for taking account of the ETS structure and efficiency, consist of market operational evaluation and economic efficiency evaluation; the former is for evaluating allowance allocation and carbon price formation, while the latter is for evaluating industrial output, number of employees, carbon leakage, quantity of carbon emissions, and impact on GDP growth.

The core of ETS is to create carbon price and apply it to economic system. While evaluating the carbon price formation, Daskalakis and Markellos [1] took the EU ETS-based carbon pricing, for instance, to analyze the influencing factors of carbon price, distributional characteristics of carbon price, and pricing of allowance derivatives. Through study of the carbon market empirical data, they concluded that carbon price is under impact from trading principals, allowance supply demand balance, and GDP growth, and they also found out that allowance allocation and carbon price are not the only factors that affect companies' investment and operation; other economic factors (e.g., capital cost and anticipated fuel price) are

© China Environment Publishing Group Co., Ltd. and Springer Nature Singapore Pte Ltd. 2019
D. Zhao et al., *A Brief Overview of China's ETS Pilots*,
https://doi.org/10.1007/978-981-13-1888-7_10

essential considerations when companies are making investment decisions. In their studies, Zhang and Wei [2] found out that financial viability, emissions reduction, and randomness are major factors that decide flow of allowances and vitality of transactions. Though carbon price is associated with the marginal cost of emissions reduction, the formation of carbon price is usually affected by multiple factors.

Through evaluation of the industrial output and trade competition in the European carbon market, Hourcad et al. [3] found out that the EU ETS has increased energy consumption cost of high-emitting companies. Electricity companies managed to pass their increased cost onto downstream sectors. Whether electricity companies are ready to try on new technologies is affected by carbon price, and mainly by electricity demand and fuel price. Through evaluation, they have found out that the free allowance allocation during EU ETS Phase 1 has brought windfall profit to certain companies, but not provided appropriate incentives in emissions reduction; in the meantime, the sectors that are exposed to fierce global competition were endowed with free allowances by the European Commission, and the ETS impact on these sectors are regularly evaluated.

Currently, the mainstream carbon trading evaluation methods refer to statistical analysis, random model simulation, partial equilibrium model, multifactor model, system dynamics model, and computable general equilibrium model. Each evaluation method has its own merits and demerits. An evaluation method is chosen on basis of the demand from both the evaluation subjects and objects. For instance, multifactor model and random model simulation need long-term data and information about trading principals and carbon prices. Such methods are apparently not fit Guangdong carbon market which is still a newborn event without sufficient trading data. Partial equilibrium model is appropriate for evaluating the emissions reduction cost of trading participants and allocation plan, but it hard to assess the impact of carbon trading policies on macroeconomy and employment, since it is unable to describe the correlations between the carbon trading participants and other entities. Statistical analysis, historical data comparison, partial case analysis, and cost-benefit analysis are able to back up the qualitative and quantitative analysis of allowance allocation and carbon price formation, and better judge the operational effect of the ETS internal modules. Computable general equilibrium (CGE) model, which is based on the input–output sheet, funds flow statement and energy balance sheet of multiple trading participants, is able to reflect the impact from changing macro-policies on micro-industrial sectors, and on sectoral carbon emissions and employment; on this basis, the economic and environmental impacts of carbon trading policies in different scenarios can be made, which will provide decision makers and stakeholders with some integral and all-round evaluation results about carbon trading policies.

CGE Model is widely used for evaluating the impact of carbon trading policies on GDP, employment, and society. Wissema and Dellink [4] analyzed the impact of carbon tax and energy tax on the Irish economy with CGE Model, and found out that carbon tax is able to significantly alter production and consumption patterns, promote adjustment to energy consumption mix, and form low-carbon economy; carbon tax and energy tax, if imposed concurrently, may generate better emissions reduction effect than single energy tax.

Edwards and Hutton [5] evaluated the UK carbon trading and allowance allocation with CGE Model, Böhringer and Welsch [6] analyzed the impact from transboundary carbon trading on global economy and trading, their analyses demonstrate that short-term emissions reduction will exert certain negative impact on GDP and industrial sectors, but the impact is not significant. When analyzing the transmission function and economic impact of the energy conservation and emission reduction policies (e.g., cap-and-trade on total emissions, carbon tax, and carbon trading), macroeconomic models are commonly used, such models mainly refer to optimization model, computable general equilibrium model, simulation model, behavioral model, data envelopment analysis (DEA) [7]. CGE Model, which is coupled with the technical models applicable to electricity sector, is used to evaluate the impact of carbon restriction policies on energy consumption and GDP growth [8]. Such model is also used for developing a special CGE Model that applies to China's energy-induced carbon emissions, in an aim to assess the complexity of the target for reducing carbon intensity and its impact on macroeconomy and industrial sectors [9], and the impact mechanism of carbon tax and ETS on supply and demand within the economic framework [10], and the impact from the adoption of new energy and nonfossil energy on the entire socio-economy and environment. The model evaluation shows that emissions reduction has different economic impact on different regions (the economic loss is usually 1–8% of GDP), but realized significant emissions cutbacks.

Overall, CGE Model is widely used for evaluating the impact from the carbon trading, energy and environmental policies on GDP growth and employment, but such evaluation at provincial level is still rare at present. Our study, in reference to the special situations in Guangdong Province, has developed a bi-regional CGE Model to evaluate the impact from carbon trading policies on industries, GDP, and employment, and also provide technical support for the quantitative evaluation of government policies and for improving policies.

10.2 Rationale of Modeling, Objectives, and Methodologies of Carbon Trading Policy Evaluation

For the purpose of evaluating the impact of Guangdong's energy and climate policies on its socio-economy, Guangzhou Institute of Energy Conversion, Chinese Academy of Sciences(GIEC, CAS), in collaboration with Japan's National Institute for Environmental Studies (NIES), developed the Guangdong–China two-regional dynamic CGE Model (ICAP-GD Model).

The process for CGE modeling is to transform the general equilibrium theory from an abstract concept into a concrete model that depicts the real economy and make it into a computable general equilibrium model. The modeling process has imported the self-optimization behaviors of all decision-making bodies like producer and those in charge of production activities, factor supply, commodity trade, and external account. The basic idea for CGE modeling is to enable producer, based on their self-optimization behaviors and under the condition of fixed commodity price and factor quantity, obtain the demand for production factors and intermediate input. It is the demand that drives producers to decide production quantity and forms social total supply. The owners of all production factors shall determine the supply based on fixed factor price, make earnings by selling factors, form the final demand for the products from all sectors, and then, together with intermediate demand, ultimately constitute total demand. Through market-based spontaneous organization, total supply will be equal to total demand.

In an abstract macroeconomic operational system, the economic agents are divided into such four sectors as residents, companies, governments, and foreign countries; and the available production factors consist of capital, labor force, and land. Residents are owner of factors and buyer of commodities; they earn income by supplying factors and confine commodity demand within the acceptable range of income. Companies are demander of production factors and supplier of commodities, while encountering a fully competitive market environment, and constraint of policies, laws, and regulations, inject companies' income into factors and intermediate products for maximizing their profit. Government is regulator of macroeconomy with fiscal revenue mainly coming from taxation; it also pays for infrastructure construction, public services, education, and other government-led undertakings. Foreign countries interact with companies through import and export, the foreign trade surplus or deficit will be saved as foreign currency earnings. Residents' savings, government's taxes, and foreign trade surplus/deficit will constitute national total savings, and be converted into next year's gross investment in form of bank loans.

ICAP-GD Model divides China into two regions (Guangdong and the other provinces/municipalities) and five modules (production sector, residents, government, foreign trade, and carbon trading), which is able to learn about the developments of Guangdong's energy, economy, and environment, and explore interprovincial trade (Guangdong and the other provinces/municipalities) and transnational trade more subtly. Such model is developed and solved by GAMS/MPSGE.

10.2.1 ICAP-GD Modeling Methodology

(1) Production sector

ICAP-GD Model involves 33 production sectors, among which there are seven energy sectors (see Table 10.1). While applying ICAP-GD Model, the activities of all production sectors are marked as CES Production Function, and then input such parameters as intermediary products, energy commodities, initial labor force, and capital. Energy commodities are made up of material and fuel.

Table 10.1 Classification of production sectors

No.	Sectors	No.	Sectors
1	Agriculture	18	Smelting and pressing of ferrous metals
2	Mining and washing of coal	19	Smelting and pressing of nonferrous metals
3	Extraction of petroleum and natural gas	20	Manufacture of metal products
4	Extraction of natural gas	21	Manufacture of purpose machinery
5	Mining and processing of ores	22	Manufacture of electrical machinery and equipment
6	Manufacture of foods	23	Production and supply of electric power and heat power
7	Manufacture of textile	24	Production and supply of gas
8	Manufacture of wood	25	Production and supply of water
9	Manufacture of paper and paper products	26	Construction
10	Other manufacturing	27	Road transport
11	Processing of petroleum	28	Railway transport
12	Coking	29	Urban public transport
13	Manufacture of raw chemical materials and chemical products	30	Water transport
14	Manufacture of cement	31	Air transport
15	Manufacture of ceramics	32	Other transport services
16	Manufacture of glass	33	Services
17	Other nonmetallic mineral products		

In reference to the input–output sheet [11] and the findings of Dai et al. [10], the electricity–heat production and supply involved in ICAP-GD Model are categorized into seven subsectors, i.e., the seven types of generating units, respectively, based on fossil fuel, natural gas, petroleum, wind and solar power, nuclear power, hydropower and garbage, biomass, and other forms of energy. These seven subsectors have their separate input–output sheet, which will be aggregated into the gross data of electricity sector, in an aim to understand the impact of new energy on Guangdong's economic development and energy consumption.

(2) Resident sector

Residents are the final consumption sector. All of the factor income, government transfer payment and revenue of residents are used for consumption or investment. If the investment amount increases at the same pace with Guangdong GDP growth over 2007–2015, then there will be consumer utility maximization under the constraint of both income level and commodity price.

(3) Government sector

Government is also the final consumption sector. Government revenue also involves taxes. Both resident and government sectors are described by CES Production Function. Government sector transfers tax revenue to provide services to the public.

(4) International trade

We assume the countries involved in ICAP-GD Model are small ones, i.e., such economy will not have any significant impact on the world economy. The energy prices on the global market have been on the rise year by year, all the other commodities have fixed base-year price, and the proportion of all products flowing on the global market is fixed.

(5) Interprovincial trade

The predominant feature of ICAP-GD Model refers to the interprovincial trade module (between Guangdong and other Chinese provinces/municipalities). The interprovincial trade is subject to Armington Assumption [10]. The products of Guangdong are separated from those of other regions, and they are described by CES. The available formulas are given as follows:

$$\text{Max}\ \pi_i = p_i \cdot Q_i - \left[p_i^{\text{md}} \cdot Q_i^{\text{md}} + \sum p_i^{\text{inf}} \cdot D_i^{\text{inf}}\right] \quad (10.1)$$

s.t.

$$Q_i = \alpha_i \cdot \left(\delta_i^{\text{md}} \cdot Q_i^{\text{md}-\rho_i} + \sum \delta_i^{\text{inf}} \cdot Q_i^{\text{inf}-\rho_i}\right)^{-\frac{1}{\rho}} \quad (10.2)$$

and

$$Q_i^{\text{md}} = \left[\frac{\alpha_i^{-\rho_i} \cdot \delta_i^{\text{md}} \cdot p_i}{p_i^{\text{md}}}\right]^{\frac{1}{1+\rho_i}} \cdot Q_i \quad (10.3)$$

$$D_i^{\text{inf}} = \left[\frac{\alpha_i^{-\rho_i} \cdot \delta_i^{\text{inf}} \cdot p_i}{p_i^{\text{inf}}}\right]^{\frac{1}{1+\rho_i}} \cdot Q_i \quad (10.4)$$

where π_i is gross profit of product i; Q_i is aggregate demand of product i; D_i^{inf} is total quantity of product i imported from other regions; p_i is price of product i; p_i^{inf} is price of product i imported from other regions; α_i is production efficiency of product i; $\delta_i^{\text{inf}}, \delta_i^{\text{md}}$, respectively, indicate the proportion of imported product i and local product $i(0 \leq \delta_i^{\text{inf}} \leq 1, 0 \leq \delta_i^{\text{md}} \leq 1, \delta_i^{\text{md}} + \sum \delta_i^{\text{inf}} = 1)$; and ρ_i denotes the CES for local and imported product i.

(6) CO_2 emissions reduction

ICAP-GD Model takes into account the CO_2 emissions from fossil fuel consumption. The emissions reduction is based on the substitution in three aspects [10]:

Fuel substitution: Along with rising carbon prices, production sector may gradually turn to use natural gas or other nonfossil energy that generates less CO_2 emissions.
Factor substitution: Owing to increasing cost in CO_2 emissions, production sector may attempt to use lower cost production factors, and attempt substitution between labor force, raw material, and capital, in order to cut CO_2 emissions.
Product substitution: Owing to rising prices for high-carbon products, residents tend to cut consumption of such products, which will indirectly lead to less CO_2 emissions from high-carbon products.

(7) Carbon trading

ICAP-GD Model sets an upper limit on carbon emissions on an exogenous basis, then figure out the profit loss from emission reduction endeavor on an endogenous basis, i.e., the emission reduction cost. When simulating carbon trading activities, there are inter-sectoral transactions, so as to balance the average marginal emission reduction cost (carbon price) of all trading sectors.

Carbon trading module is represented by the cap-and-trade system and transaction system. ICAP-GD Model assumes that the government transfers the income from selling allowances to resident sector. The curves in Fig. 10.1 depict the allowance demand from sectors C1 and C2; the axis (x) denotes the quantity of allowances to the sector, and the axis (y) indicates the marginal emission reduction cost. When carbon trading is allowed, C1 tends to purchase the allowances of ΔQ_1 in carbon market, while C2 tends to sell the allowances of ΔQ_2. ICAP-GD Model is able to discover new balance points for market clearing. See the following Formula 10.5 and 10.6:

$$\sum_{\mathrm{s}} \Delta Q_{\mathrm{s}} = \sum_{\mathrm{b}} \Delta Q_{\mathrm{b}} \tag{10.5}$$

$$\sum_{\mathrm{s}} \int_{P_{\mathrm{s}}}^{P'} C_{\mathrm{s}}(p)\mathrm{d}p = \sum_{\mathrm{b}} \int_{P'}^{P_{\mathrm{b}}} C_{\mathrm{b}}(p)\mathrm{d}p \tag{10.6}$$

where s and b, respectively, denote the seller and buyer in carbon market; ΔQ is sectoral carbon trading quantity; $C(p)$ is sectoral demand (for allowances) function; P is sectoral marginal emission reduction cost; $S = \int_P^{P'} C(p)\mathrm{d}p$ represents government transfer payment (from selling allowances) to resident sector (Fig. 10.2).

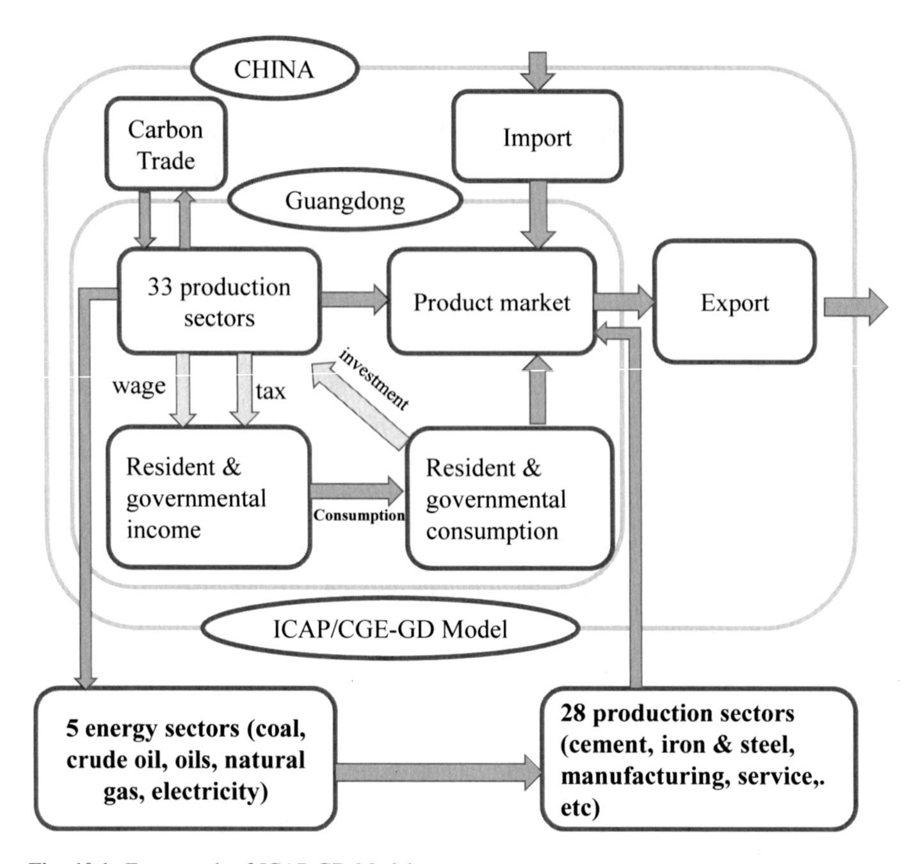

Fig. 10.1 Framework of ICAP-GD Model

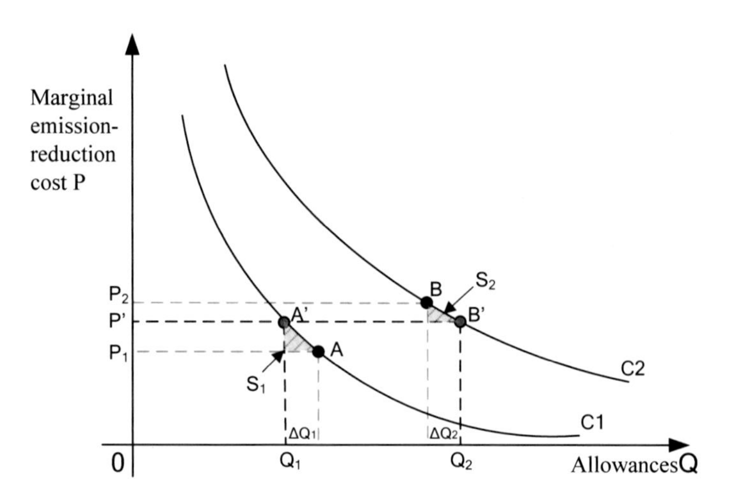

Fig. 10.2 Mechanism of inter-sectoral carbon trading [12]

10.2.2 Data Sources

(1) 2007 Input–Output Table

Input–Output (IO) Table is an important data source for constructing Social Accounting Matrix (SAM). In this study, we used China IO Table and Guangdong IO Table [11] (both the latest versions in 2007), but redivided the sectors therein. In order to conform to the four carbon trading sectors (electricity, cement, petrochemical, and iron and steel) of Guangdong Province, we separated out oil and natural gas extracting sectors, merged certain sectors, and then built the SAM with cross-entropy method.

(2) Energy balance sheet

In ICAP-GDE Model, the energy consumption and CO_2 emissions of each sector are measured by magnitude of value, but the actual energy consumption and CO_2 emissions are measured by physical quantity. In order to convert the magnitude of value into physical quantity, we shall refer to the physical quantity of energy consumption in energy balance sheet. Although the sectoral energy consumption in energy balance sheet is inconsistent with the data in IO Table, the average prices for different energy based on current year's aggregate energy consumption and magnitude of value are consistent. As such, we calculate current year's average price of each energy at first, and then convert them to energy consumption and CO_2 emissions of each sector in reference to IO Table.

Owing to disparate statistics caliber, the sectoral energy consumption in Energy Balance Sheet may vary from that in IO Table, it is generally believed that the physical quantity in Energy Balance Sheet is more reliable. In order to resolve data inconsistency, we referred to *Guangdong Statistical Yearbook* [13] which partitions the sectoral energy consumption of Guangdong, and merged and decomposed the energy consumption data of 42 sectors, so as to tally with the industrial categorization in Guangdong IO Table, and then adjusted and rebalanced the IO Table with Least Square Method and cross-entropy method. Such adjustment is of extreme importance for simulating sectoral energy consumption, CO_2 emissions and carbon trading of Guangdong.

(3) CO_2 emissions coefficient

CO_2 emission factor is computed with the IPCC recommended method [14]; specifically, CO_2 emissions quantity = fossil energy consumption quantity × emission factor × carbon oxidation rate. In a word, CO_2 emissions quantity is product of energy consumption data multiplied by coefficient (i.e., emission factor). See Formula 10.7:

$$Q_e = A_d \times E_f \times O_f \tag{10.7}$$

where

Q_e sectoral CO_2 emissions quantity measured by the unit of t-CO_2;
A_d activities that generate CO_2 emissions measured by the unit of t or 10^4 m^3;
E_f emission factor that measured by the unit of t-CO_2/t or t-$CO^2/10^4$ m^3. Calculation of E_f is based on carbon content of fuel or material, or through actual measurement or quoted from the IPCC proposed reference value.
Q_f carbon oxidation rate (90–96% for solids, 96–98% for liquids, and 100% for gases).

10.3 Evaluation Result of Guangdong ETS on Economy

10.3.1 Setting Basic Parameters

The basic data involved in ICAP-GD Model stem from China IO Table and Guangdong IO Table in 2007. We separated out several sectors in the IO Table and merged all of them into 33 sectors. The electricity sector is further divided into seven subsectors that, respectively, rely on fossil fuel, natural gas, petroleum, wind and solar power, nuclear power, hydropower and garbage, biomass, and other energy. The sectoral energy consumption by type is set in light of Guangdong Energy Balance Sheet. This model sets the basic parameters for Guangdong's macroeconomy by integrating with the provincial characteristics, the available national economic development plan, energy development plan, and industry development plan.

(1) Investment amount and population growth

In ICAP-GD Model, the investment amount is set for the entire society in light of the specific situations of Guangdong. In reference to *Guangdong Statistical Yearbook* [13] in past years, we defined the total investment amount and GDP growth rate, and set the GDP growth rate for Guangdong over 2007–2020. In reference to Guangdong national economic development plan, we set the growth rate of the provincial population over 2012–2020. The specific parameters are listed in Table 10.2.

Table 10.2 Parameters for Guangdong GDP and population growth rate

Growth rate	GDP (%)	Population (5)
2008	10.4	2.00
2009	9.7	5.11
2010	12.4	3.07
2011	10	0.61
2012–2015	8	0.64
2016–2020	7.5	0.47

(2) Clean energy-based installed capacity

Since ICAP-GD Model further divides electricity sector into seven subsectors, we shall predict and set parameters for Guangdong installed capacity and electricity output in the future. In light of the *Energy Development 12th Five-Year Plan of Guangdong Province* [15] and the *Implementation plan for accelerating the construction of clean energy in Guangdong province* [16], the model projects Guangdong installed capacity and electricity output by energy type in 2020 and defines the structure of electricity sector based on clean energy. All of the petroleum-based generators tend to be gradually phased out since 2012. The specific parameters are shown in Table 10.3.

(3) Total factor productivity and energy efficiency index

Total Factor Productivity (TFP), which is usually called Technical Progress Rate, represents the ratio between output and total factors. TFP is generated by technical progress, organizational innovation, professionalism, and production innovation. TFP indicates the extra part that the production growth rate exceeds factor growth rate.

Energy Efficiency Index (EEI) simply denotes the energy consumption for per unit of GDP or product. Therefore, the higher the energy consumption is, the lower the EEI will be. ICAP-GD Model decomposes EEI into four different types of energy ("S" for solid energy, "L" for liquid energy, "G" for gaseous energy, "E" for electricity). The TFP and EEI parameters in the model are shown in Table 10.4.

10.3.2 Setting Scenarios

(1) Base scenario

The year 2007 is defined as the base year for ICAP-GD Model. All of the parameters for GDP growth rate, population growth rate, and new energy-based

Table 10.3 Guangdong clean energy-based electricity sector in 2020

Installed capacity by energy type (Unit: GW)	Natural gas	Nuclear power	Hydropower	Wind power	Biomass and others
2007	5.2	3.8	7.7	0.3	13.4
2015	15.5	8.3	8.0	4.0	37.8
2020	22.0	18.7	8.0	7.0	40.7

Table 10.4 TFP and EEI parameters of Guangdong

	TFP (%)	EEI-S (%)	EEI-L (%)	EEI-G (%)	EEI-E (%)
2007–2012	4.0	4	3	2	2
2012–2020	3.5	4	3	2	2

electricity output are set in reference to *Guangdong Statistical Yearbook* [13] *and China* Electric Power Yearbook [17].

The parameter for new energy-based installed capacity is set in reference to *Energy Development 12th Five-Year Plan of Guangdong Province* [18], on this basis, the installed capacity and electricity output based on all types of energy over 2014–2020 are projected. TFP is set at 5% annually. Under BAU scenario, the electricity sector by 2020 will develop in line with the historical trend, and the total installed capacity will reach 132.21 GW, including 81 GW coal-based electricity, 18.41 GW gas-based electricity, 13.97 GW nuclear power-based electricity, 14.53 GW (including pumped) hydropower-based electricity, 2.50 GW wind power-based electricity, 1 GW photovoltaic electricity, and 0.80 GW biomass and garbage-based electricity.

(2) Policy scenario

Policy scenario uses the same parameters (for population, investment, EEI, and TFP) as BAU scenario, but adds two modules as carbon emissions constraint and carbon trading.

SAV scenario simulates the carbon trading of four sectors (electricity, cement, petrochemical, and iron and steel). The *Carbon Emissions Allowance Administration and Implementation Rules in Guangdong Province* (for trial) [19] set upper limit on the emissions from these four sectors. The combined emissions from the four sectors of Guangdong reached 350 Mt in 2012, while the combined emissions from the newly operated projects in these four sectors reached 59 Mt in 2015. On this basis, Guangdong set an upper limit on their combined emissions in 2013, and established the target for the provincial total carbon intensity in 2015 (to drop 19.5% below 2010 levels) and 2020 (to drop 34% below 2010 levels); in other words, Guangdong intends to cut 2020 carbon intensity by 40–45% from 2005 levels. In contrast, SAVET scenario not only sets an upper limit on total emissions, but allows the four sectors to take part in carbon trading.

LCE Scenario and LCET Scenario set a stricter upper limit on carbon emissions, manifesting the principle of the *Copenhagen Accord* that developed regions are obliged to cut more emissions, Guangdong is just doing so. From 2013 to 2015, the newly operated projects in the four sectors had 19 Mt of allowances, their carbon intensity in 2015 dropped 20.5% from 2010 levels, and the carbon intensity in 2020 may drop 19% from 2010 levels, i.e., Guangdong targets to cut 2020 carbon intensity by 45–50% from 2005 levels. LCE Scenario sets an upper limit on emissions of each sector, while LCET Scenario not only sets an upper limit on total emissions, but allows the four sectors to take part in carbon trading.

10.3.3 *Analysis of Simulation Results*

With ICAP-GD Model, the research group estimated total CO_2 emissions, sectoral CO_2 emissions, GDP growth, and variation trend of emissions reduction cost in both BAU and SAV scenarios. Based on the simulation results, they assessed of the impact of ETS on Guangdong's energy, economy and environment.

(1) CO_2 emissions quantity and intensity

In Base Scenario, the total CO_2 emissions of Guangdong in 2020 will be at least two times above 2007 levels; at that time, the emissions of Guangdong will be merely around 7.4% of the national total emissions, higher than the percentage of 7% in 2007. Moreover, Guangdong's GDP will hold around 12.7% of China's aggregate GDP in 2020, also higher than the percentage of 11.9% in 2007. The carbon intensity of Guangdong in 2017 was around 2.35 $tCO_2/10^4$ yuan, which was about 40% lower than China's average level at 3.88 $tCO_2/10^3$ US$, indicating that the technical strength, economic structure, and energy utilization efficiency of Guangdong are more advanced than China's average level. It is projected that Guangdong's carbon intensity will fall to 1.68 $tCO_2/10^4$ yuan by 2020, registering a drop of 28% from 2007 levels, yet far below the reduction of 40–45% which China pledged to achieve by 2020 in Copenhagen Summit in 2009. As such, more stringent energy and climate policies shall be unveiled to restrict the emissions of Guangdong. As demonstrated by the simulation results, if the emissions from the four energy-intensive industrial sectors (cement, petrochemical, iron and steel, and electricity) are constrained, their emissions will remarkably decrease. In SAV and LCE scenarios, the carbon intensity of Guangdong in 2020 declines 38 and 44%, respectively, from 2007 levels; and Guangdong will see its CO_2 emissions enter into a peak time in around 2020 in LCE Scenario (Fig. 10.3).

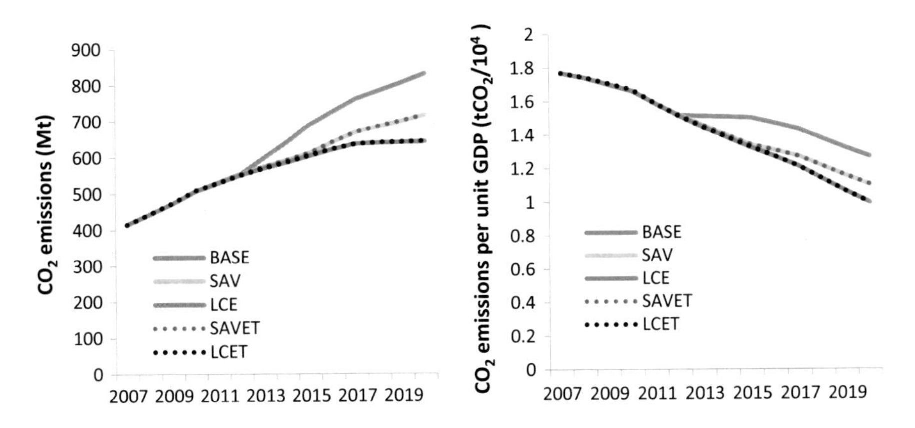

Fig. 10.3 Guangdong carbon emissions quantity and intensity [12]

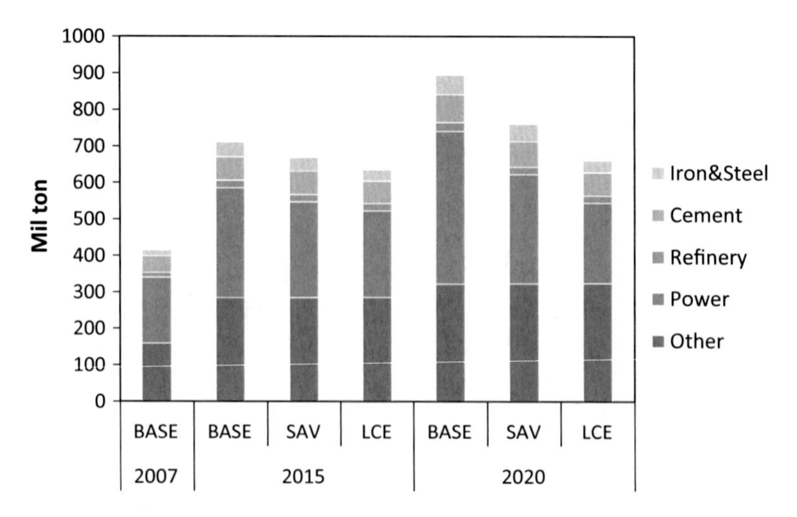

Fig. 10.4 Simulated CO_2 emissions of Guangdong's four sectors in different scenarios [12]

The changes in CO_2 emissions from the four sectors in different scenarios are shown in Fig. 10.4. By 2020, the combined emissions of the four sectors—holding around 62% of Guangdong's total emissions—will be decomposed into electricity (44.8%), cement (9.4%), iron and steel (5%) and petrochemical (2.9%), which explains the reasons why Guangdong includes the four sectors into the ETS regulation.

By 2020, the emissions reduction in electricity, cement, iron and steel, and petrochemical sectors in SAV Scenario will exceed their reduction in Base Scenario by 27, 5, 3 and 26%, respectively.

By 2020, the emissions reduction in electricity, cement, iron and steel, and petrochemical sectors in LCE Scenario will exceed their reduction in Base Scenario by 40, 23, 23, and 40%, respectively.

(2) Emissions reduction cost

When an economic system is forced to reduce CO_2 emissions, a "shadow carbon price" will arise within CGE Model. Such price may be deemed as the marginal emissions reduction cost. Owing to existence of such cost, both sectors and consumers will have extra expenses for investing the technologies and equipment that are more efficient and energy saving. In SAV Scenario, the CO_2 emissions quantity of iron and steel, petrochemical, and cement sectors will be, respectively, lower by 26, 3, and 5% than Base Scenario, and their corresponding emissions reduction cost will be 1049, 126, and 42 yuan/tCO_2. In LCE Scenario, the four sectors will be subject to more stringent emissions restrictions and face much higher carbon prices.

Carbon price is also impacted by substitution of low-carbon energy. Electricity sector is the easiest to be impacted by this factor, since massive renewable energy is available for electricity generation, which will remarkably cut the emissions reduction cost for electricity generation. In SAV Scenario, the emissions reduction rate of electricity sector is as high as 27%, ranking first among the four sectors, but the emissions reduction cost is fairly low at 121 yuan/tCO_2. While taking account of all factors, the carbon prices of cement, electricity, petrochemical, and iron and steel sectors are hereby ranked in a low-to-high order.

(3) Carbon price and trading volume under carbon trading

The economic theory manifests that the two commodities, despite the same category, may have different prices when traded in different sectors, they still have different prices. Caron emissions allowance is just a virtual product in CGE Model. Figure 10.5 shows that cement sector has the lowest carbon price, and electricity sector has the second lowest price, while petrochemical and iron and steel sectors have relatively higher carbon prices. Therefore, under most circumstances, petrochemical and iron and steel sectors are more likely to become potential buyers of emissions allowances, while electricity and cement sectors are possible to become sellers.

The SAV and LCE scenarios in Fig. 10.6 show that the cement sector will be major allowance buyer over 2013–2020. It sold about 13 MtCO_2 in 2013 and is projected to sell 5 MtCO_2 in 2020. The role that other sectors are playing in carbon trading relies on their relative carbon prices.

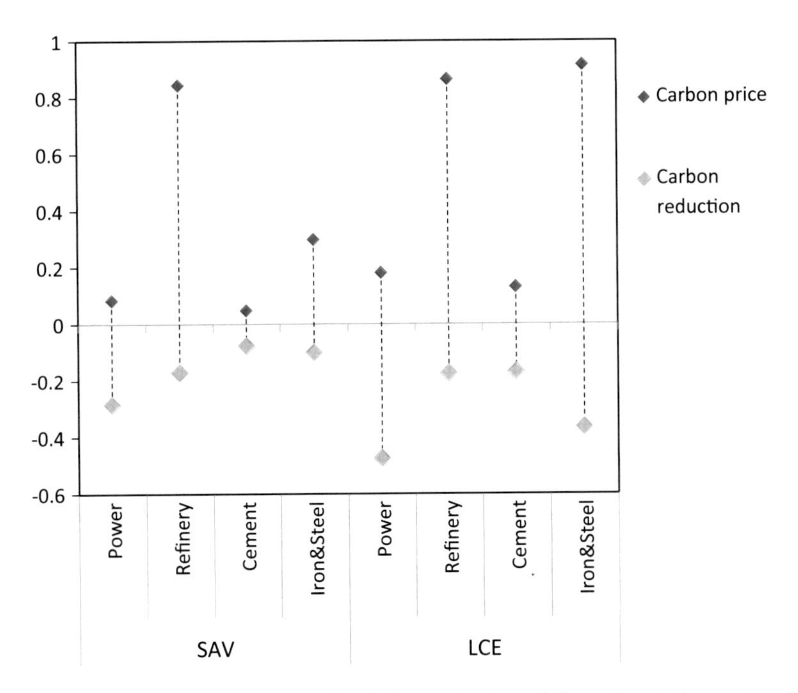

Fig. 10.5 Carbon prices and quantity of emissions reduction of Guangdong's four sectors [12]

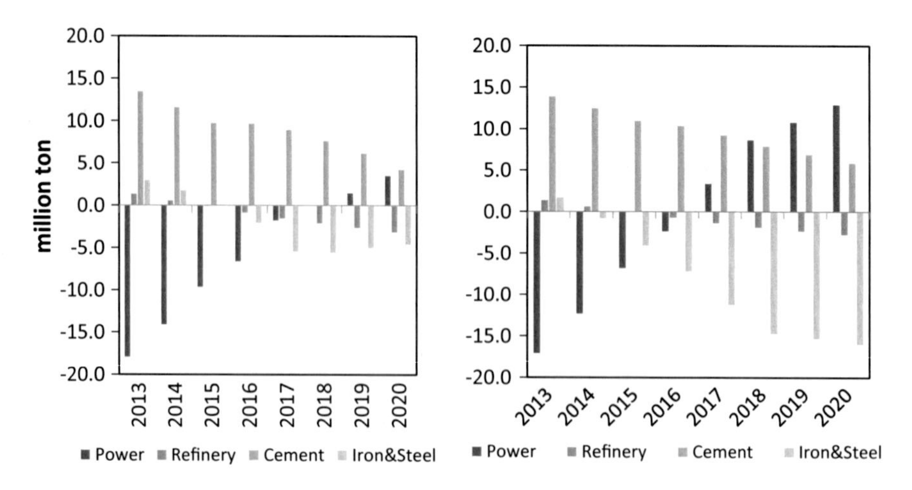

Fig. 10.6 Carbon trading volume in SAV and LCE scenarios of Guangdong's four sectors [12]

Electricity sector will be major allowance buyer from 2013 to 2016, but after that, the massive application of nonfossil energy will pull down the carbon price in electricity sector from the average level of the four sectors, mean thing that electricity sector will turn into allowance seller after 2016. Moreover, petrochemical and iron and steel sectors will replace electricity sector to become allowance buyer since 2016.

(4) Impact on employment

ICAP-GD Model is also applicable to analyze the impact of emissions restriction and carbon trading on employment. In this study, the number of employees in different sectors is calculated via Formula 10.8:

$$\mathrm{EMP}_{r,i,t} = \mathrm{SAM}_{r,i,\mathrm{lab},t} / \mathrm{SAM}_{r,i,\mathrm{lab},2007} * \mathrm{EMP}_{r,i,2007} \qquad (10.8)$$

where $\mathrm{EMP}_{r,i,t}$ represents the number of employees of sector i in area r in year t; $\mathrm{SAM}_{r,i,\mathrm{lab},t}$ denotes the labor input of sector i in area r in year t. The data in 2007 are based on the IO Table of Guangdong in 2007.

The calculation result shows that all of the four sectors joined in carbon trading are not labor-intensive sectors. In 2007, they only created 377,000 job opportunities, which were only 0.7% of the total job positions of Guangdong Province. However, the carbon trading policy that is applicable to these four sectors will impact the entire labor market (Fig. 10.7). The number of job opportunities mainly created by the six sectors is agriculture (46,103), iron and steel (41,854), service (8500), textile (3185), electronic device (2886), and metalware (2692). On the contrary, the top five sectors that have massive personnel downsizing are manufacturing (−17,553), mechanical manufacturing (−13,851), construction (−13,336), paper-making (−11,712), and chemical (−5575). The above analysis highlights that

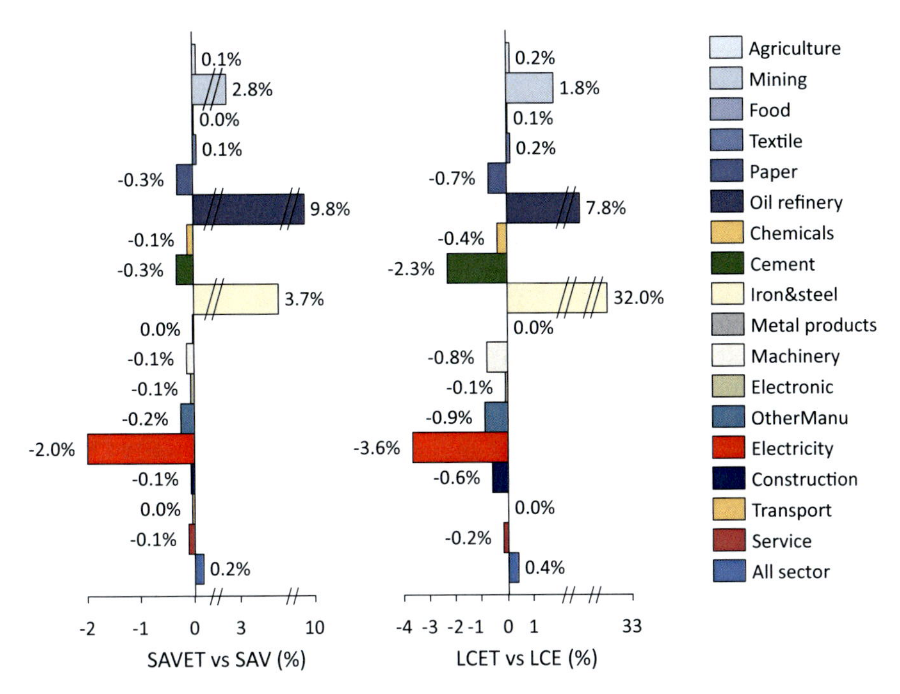

Fig. 10.7 Changes on number of employees after the sectors join in carbon trading [12]

the government shall do a conscientious job in resolving the issue of unemployment.

(5) Impact on GDP

Since both the overall and sectoral targets for cutting carbon intensity are under stricter constraint, the aggregate GDP in both SAV and LCE scenarios declines from the GDP in Base Scenario, registering a drop of 0.9 and 1.4%, respectively; which demonstrates that stricter target for emissions cutbacks will bring along more GDP losses. Therefore, the government shall consider how to design more appropriate climate policies with market as basic means, so as to mitigate GDP losses.

After comparing the carbon trading in both SAV and LCE scenarios, the research group found that the GDP losses are notably relieved by 0.8 and 1.1%, respectively, which is equivalent to saving dozens of billions of RMB. Similarly, the carbon trading will minimize the losses from emissions reduction, which helps in lowering the emissions reduction cost of the entire macroeconomy (Fig. 10.8).

(6) Sensitivity analysis

While taking account of TFP and EEI of Guangdong (GD) and other regions of China (ROC), when elasticity of substitution (σ) among products of GD changes by

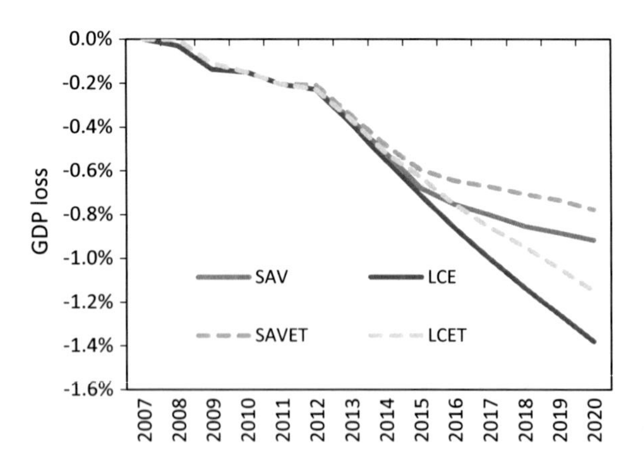

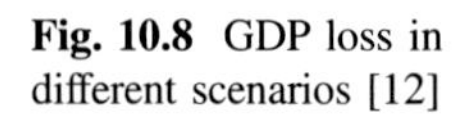
Fig. 10.8 GDP loss in different scenarios [12]

±20%, a sensitivity analysis is made for the impact of these changes on the simulated GDP loss, carbon price, trading volume, emissions reduction cost.

The result of sensitivity analysis shows that whenever there are changes in TFP and EEI of ROC, the GDP of GD will float in range of (+0.05%, −0.02%), and the carbon emissions amount will change slightly (−0.1%, +0.1%); the changes on GD's three indicators are fairly significant: GDP loss (+5.4%, −5.1%), carbon price (+8.4%, −8.7%), and carbon trading volume (+10%, −11.3%). The simulation results show that the changing economic parameters (TFP or EEI) of ROC exert a notable impact on the carbon emissions and economic situation of GD, and also influence GD's emissions reduction cost, because of the close economic ties linking GD and ROC. By means of altering TFP and EEI of GD, its total carbon emissions amount will change slightly (+0.2%, −0.2%), but its GDP will change remarkably (+12.8%, −11.6%), GDP loss (−5.8%, +6.2%), carbon price (+10.2%, −11.6%) and carbon trading volume (+13.4%, −1.6%) all change to different extent. The changes on σ will impact GDP of GD (+0.2%, −0.2%), such impact on carbon emissions (+0.03%, −0.04%) and trading volume (+0.4%, −0.6%) are moderate, but the impact on GDP loss (+3.8%, −3.8%), and carbon price (+4.9%, −5.4%) are fairly remarkable. The result of sensitivity analysis shows that the changes on TFP and EEI of ROC have a significant impact on GD's emissions reduction and GDP, while TFP, EEI, and σ also have a notable impact on GD's GDP, carbon price, and trading volume. A comparison of the two consequences shows that the TFP and EEI of GD have more significant impact on its emissions reduction and GDP (Table 10.5).

Table 10.5 Sensitivity analysis of GD relevant parameters

Sensitivity analysis	GDP (%)	Emissions (%)	GDP deficit (%)	Carbon price (%)	Trading quantity (%)
ROC+20%	0.05	−0.1	5.4	8.4	10.0
ROC−20%	−0.2	0.1	−5.1	−8.7	−11.3
GD+20%	12.8	0.2	−5.8	10.2	13.4
GD−20%	−11.6	−0.2	6.2	−11.6	−1.6
GD ES+20%	0.2	0.03	3.8	4.9	−0.4
GD ES−20%	−0.2	−0.04	−3.8	−5.4	0.6

10.4 Evaluation Conclusions and Suggestions Based on ICAP-GD Model

The ETS evaluation conclusions based on ICAP-GD Model are shown as follows:

(1) Owing to the target to cut carbon intensity by 19.5%, Guangdong carbon emissions are projected to reach 646 $MtCO_2$ in 2015 and 754 $MtCO_2$ in 2020, respectively, lower by 2 and 5% from Base Scenario. In addition to setting a cap on total emissions, Guangdong will continue carrying out the carbon trading policies, so its emissions will further fall to 644 Mt in 2015 and 728 Mt in 2020, respectively, lower 1 and 3% from Base Scenario. In a word, the ETS increases emissions cutbacks by 2 and 26 $MtCO_2$.
(2) In order to fulfill the target to cut carbon intensity by 19.5%, the economic system of Guangdong will bear an average emissions reduction cost of 18 yuan/tCO_2. Based on a cap on total emissions, the four sectors will have to bear an average cost of 129 yuan/tCO_2 to reduce carbon emissions. In contrast, in addition to the cap on total emissions, the carbon trading will notably reduce such cost to 38 yuan/tCO_2. It can be seen that the ETS boasts notable cost-effectiveness under the condition that the four sectors are bound to cut the same amount of emissions.
(3) In order to fulfill the target to cut carbon intensity by 19.5%, the GDP loss of Guangdong inflicted by the carbon trading policies is estimated at 0.956% of the aggregate GDP in 2020. Based on a cap on total emissions, there will be a GDP loss at 1.106%. Based on the scenario where the four sectors are about to cut the same amount of emissions, the practice of carbon trading is able to save economic cost and mitigate GDP loss by about 99.1 billion yuan.
(4) The carbon trading scenario is able to increase 100,000 new job opportunities in contrast to Base Scenario. These jobs are mainly distributed in service, agriculture, and mechanic manufacturing sectors, but may exert a negative impact on cement, extraction, and transportation sectors.

Based on the above analysis, the research group hereby presents the following suggestions:

(1) Cap on total emissions and carbon trading policy give rise to carbon prices, allowance transaction, compliance, and clearance by regulated companies promote carbon price conduction among different sectors. The price conduction mechanism shall be in line with the objectives of the policy design, so as to avert form market malfunction resulting from loose policy constraint on emissions reduction or the circumstance where carbon prices are unable to incentivize companies to cut emissions and invest low-carbon technologies. Therefore, relevant and detailed administration rules shall come out to deal with complicated economic situations and changes on carbon market, form a valid carbon market and standardize its operation, and facilitate the prices to be conducted to the competent authorities, companies, and other participants.

Carry out emissions reduction practice and actions, and gradually form a low-carbon industry and economy.

(2) The constraint of carbon emissions is of certain impact on energy-intensive sectors, e.g., impact the employment in these sectors. The government shall strengthen guidance and administration of labor forces, introduce the policies, and measures about personnel education and training, guide the labor forces to tertiary industry and mitigate the impact of carbon emissions reduction on manufacturing sector and the entire society.

By applying ICAP-GD Model and from the perspective of general equilibrium economic system, the research group analyzed the implications of the carbon trading policies on macroeconomy and the influencing factors, and made a quantitative evaluation of the impact. The model construction is based on the assumption of a full market competition. The carbon prices are formed in transparent information of market players and complete emissions reduction potentials.

Therefore, the assumptions may vary from the actual situations of Guangdong carbon market. The simulated carbon price, trading volume, and emissions reduction cost are related to the economic activities and projected emissions quantity, which will bring some uncertainties to the simulation result. The elasticity parameters in the model need further validation and improvement from the actual carbon market data, in order to support policy-makers to make decisions.

References

1. Daskalakis, G., Markellos, R.N., 2008. Are the European carbon markets efficient. Review of Futures Markets 17, 103–128.
2. Yuejun Zhang, Yiming Wei, Interpreting the mean reversion of international carbon future price: empirical evidence from the EU ETS [M] System Engineering Theory and Practice, 2011, 31, 214–220.
3. Hourcade, J.-C., Demailly, D., Neuhoff, K., Sato, M., 2002. Climate Strategies Report: Differentiation and Dynamics of EU ETS Industrial. IMF Staff papers 16, 159–178.
4. Wissema, W., Dellink, R., 2010. AGE assessment of interactions between climate change policy instruments and pre-existing taxes: the case of Ireland. International Journal of Global Environmental Issues 10, 46–62.
5. Edwards, T.H., Hutton, J.P., 2001. Allocation of carbon permits within a country: a general equilibrium analysis of the United Kingdom. Energy Economics 23, 371–386.
6. Böhringer, C., Welsch, H., 2004. Contraction and convergence of carbon emissions: an intertemporal multi-region CGE analysis. Journal of Policy Modeling 26, 21–39.
7. Yiming Wei, Zhifu Mi, Hao Zhang. Review on climate policy modeling: An analysis based on bibliometrics method [J] Advances in Earth Science. 2013, 28, (8) 930–938.
8. Ruamsuke, K., Dhakal, S., Marpaung, C.O., 2015. Energy and economic impacts of the global climate change policy on Southeast Asian countries: a general equilibrium analysis. Energy 81, 446–461.
9. Can Wang, 2003. Climate change policy simulation and uncertainty analysis: a dynamic CGE model of China [D] Beijing: Tsinghua University, 2003.

10. Dai, H., Masui, T., Matsuoka, Y., Fujimori, S., 2011. Assessment of China's climate commitment and non-fossil energy plan towards 2020 using hybrid AIM/CGE model. Energy Policy 39, 2875–2887.
11. Department of National Accounts of National Bureau of Statistics. IO Table-China-2007 [M]. Beijing: China Statistics Press, 2009.
12. Wang P, Dai H C, Ren S Y, et al. Achieving Copenhagen target through carbon emission trading: Economic impacts assessment in Guangdong Province of China[J]. Energy, 2015, 79 (79):212–227.
13. Statistics Bureau of Guangdong Province. Guangdong Statistical Yearbook [M]. Beijing: China Statistics Press, 2008-2013.
14. Ministry of Science and Technology, Economy and Energy. Supply of an instrument for estimating the emissions of green house effect gases coupled with the energy matrix [R/OL]. 2010 [2014-06-11]. http://ecen.com/matriz/eee24/coefycin.htm.
15. Guangdong Provincial Development and Reform Commission. Twelfth Five-Year Plan for National Economic and Social Development of Guangdong Province [EB/OL]. 2011 [2014-07-01]. http://www.gddpc.gov.cn/fgzl/fzgh/ztgh/sewghgy/201106/t20110615_155249.htm.
16. Guangdong Provincial Development and Reform Commission. Implementation plan for accelerating the construction of clean energy in Guangdong province [EB/OL].2015 [2015-07-13]. http://zwgk.gd.gov.cn/006939756/201507/t20150727_594912.html.
17. State Grid Corporation of China et al. China Electric Power Yearbook [M]. Beijing: China Electric Power Press, 2008-2013.
18. Guangdong Provincial Development and Reform Commission. Energy Development 12th Five-Year Plan of Guangdong Province [EB/OL].2013 [2014-07-01]. http://www.gddpc.gov.cn/fgzl/fzgh/zxgh/sewzx/201309/t20130926_220199.htm.
19. Guangdong Provincial Development and Reform Commission. Notice on Carrying out the Work for Guangdong First Emissions Allowance Allocation (for trial) [EB/OL]. 2013 [2014-07-01]. http://www.gddpc.gov.cn/xxgk/tztg/201311/t20131126_230325.htm.

Appendix

Notice of People's Government of Guangdong on Issuing the Plan for Carrying out the Work about the Carbon Emissions Trading Pilot Program in Guangdong Province

GD GOV [2012] No. 264

All municipal people's governments at and above the prefecture level, all county (city or district) people's governments, and all of the departments of and directly under the provincial government:

The *Plan for Carrying out the Work about the Carbon Emissions Trading Pilot Program in Guangdong Province* is hereby issued for your conscientious implementation. In case any problem takes place while executing the plan, please notify the provincial development and reform commission of the matter.

September 7, 2012

People's Government of Guangdong

Plan for Carrying out the Work about the Carbon Emissions Trading Pilot Program in Guangdong Province

In light of the State Council's *Work Plan for Greenhouse Gas Emissions Control during the 12th Five-Year Plan Period* and the *Twelfth Five-Year Plan for National Economic and Social Development of Guangdong Province*, as well as the work deployment of the National Development and Reform Commission in carrying out the carbon emissions trading pilot program, the *Plan for Carrying out the Work about the Carbon Emissions Trading Pilot Program in Guangdong Province* is hereby formulated to conscientiously implement this pilot program in Guangdong

© China Environment Publishing Group Co., Ltd. and Springer Nature Singapore Pte Ltd. 2019
D. Zhao et al., *A Brief Overview of China's ETS Pilots*,
https://doi.org/10.1007/978-981-13-1888-7

Province and fulfill the target in controlling greenhouse gas emissions based on market mechanism and at a lower cost.

1. **Guiding Ideology and Work Objectives**

(1) **Guiding ideology**

Center on the core tasks for expediting industry transition and upgrading, constructing a happy Guangdong, strengthening government guidance and market operation, and further raising the awareness of companies and all sectors of society of the importance in controlling greenhouse gas (GHG) emissions. Give full play to the role of market mechanism in fulfilling such objectives as energy conservation, carbon emissions cutbacks, and control of aggregate energy consumption during the 12th Five-Year-Plan period (2011–2015). Make innovations in the institutional mechanisms for promoting industry transition and upgrading and balancing development between regions. Learn from the carbon emissions trading experiences both home and abroad, take account of the specific situations of Guangdong Province, improve the system design of an emissions trading mechanism, and make explorations for building a nationwide carbon market in the future.

(2) **Work Objectives**

By 2015, Guangdong shall have basically built an administration system for reasonable allocation of carbon emissions credit between market principals and between regions, and initially formed a carbon emissions trading scheme (ETS) that is based on the provincial situations, and characterized with a well-functioned structure, standardized administration and sound operation, and ranking a leading place among the regional carbon markets.

Guangdong ETS shall constantly develop and improve, and an interprovincial ETS shall take shape by 2020.

2. **Overall Arrangement**

(1) **Trading products**

The trading products in Guangdong carbon market mainly refer to the emissions allowances, i.e., the government will allocate quantified CO_2 emissions credit to companies and other trading principals. After filed with by the national or provincial development and reform commission, the China Certified Emission Reduction (CCER) is allowed to be traded as a supplementary product. Explorations in innovating carbon trading products are encouraged.

(2) **Trading principals**

The trading principals in Guangdong carbon market refer to the companies that are involved in the ETS coverage scope (hereinafter referred to as the "regulated companies"); yet active explorations are also made in drawing in investment institutions, and other market players into the ETS regulation. The government allocates emissions allowances to the covered enterprises and exercises supervision and administration of these companies which are obliged to cut CO_2 emissions in line with the quantity of allowances, and they are allowed to gain economic benefit or emissions rights and interests.

(3) **Trading platform**

China (Guangzhou) Emissions Exchange (CGEE) is a designated carbon emissions trading platform for Guangdong carbon market, i.e., all of the provincial emissions trading activities shall be based on the platform.

(4) **Phases of ETS implementation**

The implementation of Guangdong ETS pilot program is split into three phases: Phase I (2012–2015) is a trial period; Phase II (2016–2020) is a period where the ETS operation will be gradually improved; Phase III (2020 onward) is a period featuring mature operation of the ETS.

3. **Major Tasks**

Gradually establish and improve a government-led mechanism for supervision and administration of carbon market principals, form a market mechanism that helps fulfill the targets in energy conservation and emissions reduction and industry restructuring, and safeguard smooth progress of carbon trading activities.

(1) **Build an emissions information reporting and verification mechanism**

i. Build an emissions information reporting system. Define the major companies that shall report their emissions information (hereinafter referred to as the "reporting companies") in reference to the scope of key energy-intensive sectors, and gradually broaden the scope of reporting companies. All of the covered enterprises are obliged to report their emissions information.
ii. Build a verification system for verifying the emissions from covered enterprises. Foster and authorize qualified third-party verification institutions to verify the carbon emissions information reported by the covered enterprises.
iii. Build an electronic information service system. Establish a corresponding electronic information system to facilitate companies' reporting and third-party institutions' verification activities.

(2) **Build an emissions allowance administration system**

i. Strengthen administration of aggregate carbon emissions. Establish the provincial and municipal (prefectural) targets for emissions cutbacks in a scientific approach in reference to year-on-year decrease in carbon intensity, year-on-year decrease in aggregate emissions and relevant constraint indicators, and integrated with the actual situations in socioeconomic development. Such targets will serve as basis for emissions allowance administration.

ii. Scientific and reasonable allocation of emissions allowances. Define the annual aggregate carbon emissions that fall under government regulation by taking full account of the socioeconomic growth trend and construction of key projects. Stipulate relevant rules, issue allowances to the covered enterprises, and include the newly built fixed-asset investment projects (at prescribed production scale) into allowance administration. Support relevant companies to establish a carbon asset administration system.
iii. Set up an emissions allowance registration system. Build a corresponding electronic system for registration of allowance account and for recording the details about allowance distribution, alteration and cancelation, etc.

(3) **Build an allowance trading operational system**

i. Standardize construction of CGEE. In accordance with the relevant national regulations, CGEE shall be built into an emissions allowance trading platform that facilitates allowance transactions within Guangdong Province and China at large.
ii. Stipulate business rules for allowance trading. Work out and constantly improve the business rules for trading matchmaking, price formation, allowance delivery, review and verification, capital settlement, information disclosure, risk control, entrustment of an agent and dispute mediation. It is based on these rules that the transaction activities are carried out. And the allowance trading activities shall follow the principles of openness, fairness, and impartiality.
iii. Set up an electronic transaction system. Build a supporting electronic information system that has the functions for online allowance bidding, opening transaction account, and recording transaction information, and it is able to create a complete hardware and software environment for carrying out allowance trading activities.
iv. Build an allowance trading supervision and administration mechanism. Strengthen supervision and administration of allowance trading process and operation of emissions exchange.

(4) **Carry out GHG voluntary emissions reduction**

Actively promote the provincial institutions, companies, groups, and individuals to join in the national GHG voluntary emissions reduction in light of the *Interim Measures for the Administration of Greenhouse Gas Voluntary Emission Reduction*.

(5) **Explore the construction of an interprovincial ETS**

Make explorations in building an interprovincial ETS together with other provinces (municipalities), strive for government sponsorship and carry out such plan when the time is ripe.

4. **Guarantee Measures**

(1) **Strengthen organizational leadership**

Under the framework of holding joint conferences on carrying out the national pilot low-carbon program in provinces, Guangdong has set up a special coordination and leading group for implementing the pilot ETS program, with the director of the provincial development and reform commission (DRC) as the chief, and the concerned officials with Guangzhou Municipal People's Government and the provincial DRC as deputy chiefs; the members of the group come from the provincial commission of economy and informatization, department of finance, department of forestry, state-owned assets supervision and administration commission, price bureau, quality and technology supervision bureau, office of legislative affairs and finance office, as well as from Guangzhou Municipal Development and Reform Commission, finance office, and CEGG. The office of the leading group is in the provincial DRC. The provincial DRC is the competent authority taking charge of the provincial carbon emissions and allowance trading, coordinating the work on the ETS. Establish a working group for ETS research and system design with the provincial experts as members and national experts as advisors. The competent institutions at both the municipal and provincial levels shall strengthen coordination, refine their work and tasks, and quicken the pace of work.

(2) **Strengthen construction of a legal system**

Stipulate the interim measures for the administration of carbon emissions trading in Guangdong Province, define the scope of carbon trading principals, and work out the regulations on emissions reporting and verification, allowance allocation, transaction institution, responsibilities of supervision and regulation. Closely watch the implementation of the ETS pilot program, draw up experiences without delay, present legislation plans for tackling climate change or carbon emissions administration, and kick off the legislative procedure at an appropriate time.

(3) **Build up construction of capabilities**

Emphasize the basic research about ETS, and constantly improve working ideas and approaches. Conduct exchanges with the foreign counterparts and learn from the advanced experiences both home and abroad. Develop the ETS consultation and verification agencies and strengthen their administration. Hold special trainings about carrying out the ETS in an extensive manner.

(4) **Increase financial support**

In light of the relevant regulations, the provincial low-carbon development fund shall be mostly diverted to the ETS institutional researches and their work systems that are qualified candidates for receiving the fund.

(5) **Increase publicity and guidance**

Broadly publicize the rationale, rules, and relevant policies and measures about the ETS, and guide companies and other market principals to actively perform their obligations in cutting GHG emissions and taking part in carbon trading.

5. **Major Arrangements in Trial Period (2012–2015)**

(1) **Scope of the companies that report carbon emissions**

The reporting companies refer to the industrial companies within the administrative area of Guangdong Province and emit 10,000 t/CO_2 and above (or consume comprehensive energy of 5000 tCO_2e) in any year over 2011–2014. The name list of these companies shall be determined by the provincial development and reform commission and competent authorities after deliberation and consideration of the ETS implementation process and characteristics of the industrial sectors. They shall organize such companies to report their carbon emissions information by stages and step by step, and conduct studies on covering the key companies in transportation and construction sectors into the scope of reporting companies.

(2) **Scope of the companies that cap aggregate emissions and join in allowance trading**

The covered enterprises refer to the industrial companies (electricity, cement, iron and steel, ceramics, petrochemical, textile, nonferrous, plastics, paper-making, etc.) within the administrative area of Guangdong Province and emit 20,000 t/CO_2 and above (or consume comprehensive energy of 10,000 tce) in any year over 2011–2014. The name list of these companies shall be determined by the provincial development and reform commission and competent authorities after deliberation and consideration of the ETS implementation process and characteristics of the industrial sectors. They shall organize such companies to cap their total emissions and join in carbon trading by stages and step by step. Efforts shall be made to draw the key companies in transportation and construction sectors to be subject to capping on total emissions and carbon trading.

(3) **Allowance allocation**

The provincial development and reform commission shall, in light of the historical CO_2 emissions (2010–2012) of the covered enterprises and the characteristics of the corresponding industrial sectors, allocate the annual emissions allowances over 2013–2015 to these companies at one time. While considering the macroeconomic situations and the corporate last year's emissions, the commission shall appropriately adjust their current year's allowances. Put in place the allowance paid-use system, and integrate free allocation and paid allocation at the initial phase of the ETS, with the former playing a leading role.

Through an evaluation of the newly built fixed-asset investment projects, if their annual comprehensive energy consumption reaches 10,000 tce and above, then the provincial development and reform commission shall grant them with free allowances or a portion of paid allowances based on the evaluation result and projected annual aggregate emissions of the province. Whether such projects are able to access to the allowances that are equal to the evaluation result may serve as the important basis for the investment authorities at all levels to go through approval formalities.

(4) **Allowance use**

The provincial development and reform commission shall, within a prescribed time span each year, deduct the allowances that are equal to the last year's actual emissions (verified) from the allowance account of the covered enterprises, in order to offset their last year's actual emissions. The covered enterprises are allowed to sell the surplus allowances or bank them for use in coming year (invalid by the 2015 compliance period). But they shall purchase additional allowances within a prescribed time limit to make up for inadequate allowances so as to perform their obligations in cutting emissions. The economic, legal, technical, and necessary administrative means shall be integrated to enhance the supervision and administration of the compliance of the regulated companies. For the owners of newly built projects, they may not trade the allowances before operation of the projects; in other words, their allowances will become tradable after the projects are put into operation in light of relevant regulations.

(5) **Complementary mechanism**

The provincial development and reform commission and the competent departments shall, in light of the special situations of Guangdong Province and the relevant national requirements, formulate the filing rules and operational approaches for the CCER from forestry carbon sequestration and similar projects. The CCER or GDCER are allowed to join in Guangdong's ETS system in light of relevant regulations.

(6) **Working progress**

The ETS pilot program of Guangdong places focuses of the Phase 1 work on testing the ETS pilot program in certain key industrial sectors. Phase I could be split into three stages:

i. Preparation stage (2012-1H2013). Kick off the trading with project-based GHG VER. Stipulate relevant specifications and business rules, establish emissions information reporting and verification systems, emissions allowance registration, emissions supervision, and administration system. Establish an official carbon emissions exchange for listed trading.
ii. Implementation stage (2H2013-2014). Kick off the allowances-based trading, and constantly improve the ETS administration and transaction systems. Conduct preliminary studies on constructing an interprovincial ETS, and strengthen coordination on the work for interprovincial ETS.
iii. Deepening stage (2015). Promote the sound progress of GHG VER trading and provincial allowance trading, and strive to be the first in kicking off the

pilot work for a constructing interprovincial ETS. Summarize and evaluate the work about carrying out the ETS pilot program, and study the working ideas and implementation plan for implementing the ETS during the 13th Five-Year-Plan period (2016–2020).

Interim Measures for Carbon Emissions Administration in Guangdong Province

Chapter 1 General Provisions

Article 1 These measures are hereby formulated, in light of the special situations of Guangdong Province, for achieving the target for greenhouse gas (GHG) emissions control by giving full play to the role of market mechanism and standardizing administration of carbon emissions activities.

Article 2 These measures are applicable to the carbon emissions reporting and verification, allowance allocation, settlement, and transaction within the administrative jurisdiction of Guangdong Province.

Article 3 The administration of carbon emissions shall abide by the principles of openness, fairness, and integrity, and insist on government guidance and market-oriented operation.

Article 4 Guangdong Provincial Development and Reform Commission takes charge of organization, implementation, coordination, and supervision of the work on provincial-wide carbon emissions administration.

All municipal people's governments at and above the prefecture level shall guide and support the companies within their jurisdiction to cooperate with them in carbon emissions administration.

All municipal development and reform commissions at and above the prefecture level shall organize the companies within their jurisdiction to perform the work in carbon emissions reporting and verification.

All provincial departments that take charge of economy and informatization, fiscal, housing and urban–rural construction, transportation, statistics, price regulation, quality supervision, or finance shall perform their own duties in carbon emissions administration.

Article 5 Develop voluntary emissions reduction (VER) projects such as forestry carbon sequestration, guide companies and institutions to adopt measures for saving energy and cutting carbon emissions, raise the public awareness of participating in emissions reduction, and promote low-carbon and energy-saving actions across the whole society.

Chapter 2 Carbon Emissions Reporting and Verification

Article 6 Guangdong Province has put in place the system for carbon emissions reporting and verification.

The industrial companies that emit carbon dioxide (CO_2) at and above 10,000 tons annually, together with the institutions (hotels, restaurants, financial, commercial, and public organs) that emit CO_2 at and above 5000 tons annually are called "regulated companies/institutions". The industrial companies that emit CO_2 more than 5000 tons but less than 10,000 tons annually are called "reporting companies".

The norms and scope of the covered enterprises/institutions in transport sector are proposed by Guangdong Provincial Development and Reform Commission in collaboration of the transport department. In light of their carbon emissions administration, they will fall into the scope of reporting and verification in batches.

Article 7 The covered enterprises/institutions shall produce carbon emissions report for the previous year as prescribed and submit the report to Guangdong Provincial Development and Reform Commission.

The covered enterprises/institutions shall entrust a verification institution to verify their carbon emissions report, cooperate with the verification institution and bear all expenses therefrom.

If the annual CO_2 emissions quantity recorded in the carbon emissions report differ from the result in the verification report by more than 10% (100,000 tons), the carbon emissions report shall be subject to a reverification organized by Guangdong Provincial Development and Reform Commission.

All municipal development and reform commissions at and above the prefecture level shall conduct a spot check of the carbon emissions report submitted by the covered enterprises, and the expenses therefrom are included into the fiscal budget at the corresponding level.

Article 8 The verification institutions that verify the carbon emissions report shall have the corresponding qualifications associated with their verification activities, and fixed premises and necessary facilities that enable them to conduct verification within Guangdong Province.

Both the institutions and personnel that undertake verification shall perform carbon emissions verification according to law and in an independent and impartial manner; take responsibility for the standardability, authenticity, and accuracy of their verification report; and perform the obligation of confidentiality according to law and bear the legal liability.

Article 9 The charging standards for verifying carbon emissions are determined by the pricing authority of Guangdong Province.

Chapter 3 Administration of Allowance Allocation

Article 10 Guangdong Province has established a carbon emissions allowance (hereinafter briefed as "allowance") administration system. The covered enterprises/institutions, together with the new entrants (including expanded or rebuilt ones) that emit CO_2 at and above 10,000 tons annually, are under allowance administration. The other emitting companies/institutions may apply to Guangdong Provincial Development and Reform Commission for joining in allowance administration.

Article 11 The total amount of allowances are defined by People's Government of Guangdong Province by taking account of the national overall target for GHG control, Guangdong's development plan of the provincial key industries and the objective for reasonable control of aggregate energy consumption.

Of the total amount of allowances, despite the most portion granted to the covered enterprises/institutions, part of them are used as reserve allowances which are made up of the allowances distributed to new entrants and those for market regulation.

Article 12 Guangdong Provincial Development and Reform Commission shall produce a provincial allowance allocation plan, and clarify the allocation principles, methodologies, and procedures, which are subject to review by the allowance allocation review board, and finally reported to People's Government of Guangdong Province for ratification and announcement.

The allowance allocation review board consists of the technical, economic, low-carbon, and energy experts from Guangdong Provincial Development and Reform Commission and other competent authorities, industry associations, and companies. The number of experts may not be less than 2/3 of the total number of board members.

Article 13 The annual allowances distributed to the covered enterprises/institutions are decided by Guangdong Provincial Development and Reform Commission with the Benchmarking and Grandfathering methodologies in light of sectoral baseline emissions, emission reduction potentials, and companies' historical emissions.

Article 14 The allowances distributed to the covered enterprises/institutions are based on free and paid allocation, with the percentage of free allowances gradually decreasing.

Guangdong Provincial Development and Reform Commission will allocate free allowances to the covered enterprises/institutions pursuant to the prescribed proportions on every July 1.

Article 15 In case any incorporation occurs among the covered enterprises/institutions, their previous allowances and concomitant rights and obligations shall be inherited by the later incorporated companies/institutions; or if any separation occurs within a regulated company/institution, a separate allowance allocation plan shall be produced and then reported to the provincial and municipal development and reform commissions to for archiving.

Article 16 If the covered enterprises/institutions are forced to halt operation/ production owing to changes on category of products and business scope, or encounter significant changes on their operation/production, they shall submit the applications to Guangdong Provincial Development and Reform Commission for redefining the amount of allowances.

Article 17 If the covered enterprises/institutions revoke their business license, halt operation/production or move out of Guangdong Province, they shall, within one month before completion of the closure or removal formalities, submit the carbon emissions report and verification report and return the remaining allowances to the issuing authority.

Article 18 The covered enterprises/institutions shall, in light of their last year's actual carbon emissions, settle the allowances and apply to Guangdong Provincial Development and Reform Commission for writing off their allowances. The annual surplus allowances may be banked for compliance in the following compliance period or used for trading.

Article 19 The covered enterprises/institutions are allowed to use the Chinese Certified Emission Reduction (CCER) to offset a portion of their actual carbon emissions; yet the percentage of CCERs may not be over 10% of the companies' last year's actual emissions, and more than 70% of the CCERs shall stem from the VER projects within Guangdong Province.

If the CCERs are generated within the emission boundary of the covered enterprises/institutions, they are not allowed to offset the carbon emissions of the covered enterprises/institutions.

One ton of the Chinese Certified Emission Reduction (CCER) in CO_2 equivalent (1 t/CO_2e) is able to offset one ton of carbon emissions.

Article 20 The allowances granted to the new entrants are first confirmed by the municipal development and reform commission at and above the prefecture level which will validate their emissions assessment results, and then determined by Guangdong Provincial Development and Reform Commission. The new entrants may not access to free allowances before they purchase a certain proportion of paid allowances.

Article 21 Guangdong Provincial Development and Reform Commission, by means of holding auctions, distributed paid allowances via the trading platform designated by People's Government of Guangdong Province regularly every year. The auction floor price is jointly determined by Guangdong Provincial Development and Reform Commission and pricing authority.

The allowances allocated via auction are granted to the existing covered enterprises/institutions and new entrants, and also reserved for market regulation.

Article 22 All the allowances in Guangdong Province are subject to registered administration. The allowance allocation, alteration, settlement, and cancelation shall be registered in the designated allowance registry according to law and comes into effect as of the date of registration.

Chapter 4 Allowance Trading Administration

Article 23 Guangdong Province is implementing the carbon emissions allowance trading system. The trading principals are made up of the covered enterprises/ institutions, new entrants, and other eligible organizations and individuals.

Article 24 The trading platform is a carbon emissions exchange (hereinafter referred to as "emissions exchange") which shall perform the following obligations:

(1) Stipulate transaction rules;
(2) Provide a transaction venue, system facilities, and services, and organize trading activities;
(3) Establish a capital settlement system, conduct transaction settlement, liquidation, and fund supervision according to law;
(4) Establish a trading information administration system, publish trading dynamics, prices and quantity, and promptly disclose the information that may lead to major market changes;
(5) Establish a risk administration system which takes charge of risk control, supervision, and administration of trading activities; and
(6) Shoulder other responsibilities that are provided for in accordance with law and relevant regulations.

The transaction rules shall not be published before examined and reviewed by the provincial development and reform commission and financial administration.

Article 25 The allowance trading is carried out via open bidding, negotiated transfer, and other standardized approaches that are permitted by law and relevant regulations.

Article 26 The allowance trading prices are determined by market participants based on the allowance supply–demand pattern. Neither entity nor individual may manipulate the trading prices via fraud, malicious collusion or other means.

Article 27 The trading participants shall pay for the trading formality fees as required. The charging standard for the formality fee is proposed by the emissions exchange, and then executed upon validation of the provincial pricing authority.

Article 28 Guangdong Province has been making explorations for building a trans-regional carbon emissions trading market, so it encourages nonlocal companies to take part in the provincial carbon trading.

Chapter 5 Supervision and Administration

Article 29 Guangdong Provincial Development and Reform Commission shall, via government website or news media, regularly disclose the information about the implementation of these measures by the covered enterprises/institutions and reporting companies.

Guangdong Provincial Development and Reform Commission shall make public the directory of verification institutions, and strengthen supervision and administration of these institutions and their verification activities.

Article 30 Guangdong Province shall establish the corporate emissions reporting and verification system and allowance transaction system. The covered enterprises/institutions and reporting companies shall, by following the instructions, open an account in corresponding systems and upload relevant data.

Article 31 If the covered enterprises/institutions disagree with the verification result about their annual actual emissions or allowance allocation, they are allowed to apply to Guangdong Provincial Development and Reform Commission for a reverification according to law. If the dissension arises from verified annual emissions, Guangdong Provincial Development and Reform Commission shall entrust a verification institution to conduct a reverification. If the dissension arises from allowance allocation, the commission shall confirm the fact and make a written reply to the applicant within 20 days.

Article 32 Guangdong Provincial Development and Reform Commission shall create credit files for the covered enterprises/institutions, verification institutions, and emissions exchanges, and record, integrate, and publish the credit-related information about their emissions administration and trading activities.

Article 33 Under the same conditions, the companies that have performed their emissions reduction obligations are prioritized to apply for the special funds, e.g., national low-carbon development fund, energy saving and emissions reduction fund, renewable energy fun, and circular economy fund. And they are among the first applicants to receive support from provincial low-carbon fiscal subsidy and other special funds for encouraging energy saving and emissions reduction and development of circular economy.

Article 34 Encourage financial institutions to explore the financing services about the products traded in carbon market and provide financing support for the companies/institutions that are subject to allowance administration.

Article 35 The revenue from selling allowances and the expenditure therefor are under separate fiscal administration.

Chapter 6 Legal Liabilities

Article 36 Any regulated company/institution or reporting company that violates Article 7 that are provided for in measures or commits one of the following actions, it shall to make a correction within a prescribed time limit as instructed by Guangdong Provincial Development and Reform Commission; otherwise, it shall be imposed a fine as follows:

(1) If a regulated company/institution makes a false report, conceals or rejects to perform the obligation of reporting its emissions, it shall be imposed a fine of no less than 10,000 yuan but no more than 30,000 yuan;
(2) If a regulated company/institution obstructs the verification institution to go for on-site verification and rejects to turn over the relevant evidence, it shall be imposed a fine of no less than 10,000 yuan but no more than 30,000 yuan, or a fine of 50,000 yuan under a serious circumstance.

Article 37 If a regulated company/institution violates Article 18 that are provided for in measures and fails to pay off the allowances, it shall be urged by Guangdong Provincial Development and Reform Commission to perform the payment obligation; otherwise, it shall pay twice more than the amount of unpaid allowances in next compliance period, and imposed a fine of 50,000 yuan.

Article 38 If the emissions exchange commits one of the following behaviors, it shall be urged by Guangdong Provincial Development and Reform Commission to make correction, and imposed a fine of no less than 10,000 yuan but no more than 50,000 yuan.

(1) Fail to make public the transaction information as required;
(2) Fail to establish and implement the risk administration system.

Article 39 In case the verification institution violates the provisions in the second paragraph of Article 8 as provided for in these measures and commits one of the following behaviors, it shall be urged by Guangdong Provincial Development and Reform Commission to make correction within a prescribed time limit, and imposed a fine of no less than 30,000 yuan but no more than 50,000 yuan.

(1) Produce a fake or false verification report;
(2) Use or publish the business secrets and emissions information of the verified company/institution without permit.

Article 40 If the development and reform commission and other competent authorities and their staff members violate the provisions as provided for in these measures and commit one of the following behaviors, they shall be urged by the superior authority or supervision agency to make correction and subject to public criticism. Under a serious circumstance, the persons in charge and persons directly responsible shall receive disciplinary sanction imposed by the department in charge of appointment and dismiss. Anyone that is suspected of committing a crime shall be transferred to the judicial organs for investigation of criminal responsibility

(1) Seek unjust benefit by taking advantage of the work related to allowance allocation, emissions verification and validation, and administration of verification institution;
(2) Fail to correct or punish the illegal activities that he/she has discovered;
(3) Disclose the confidential information about allowance trading in violation of the relevant regulations and thereby cause serious consequences; and
(4) Commit any other violations like abuse of power, neglect of duty or irregularities for favoritism.

Chapter 7 Supplementary Provisions

Article 41 The specific provisions about the verification of corporate carbon emissions report, allowance allocation, and financial services are separately provided by the provincial development and reform commission and financial authority in reference to the provisions of these measures.

Article 42 Definitions of the terminology in these measures:

(1) *Carbon emissions allowance* is quantified emissions credit which is issued by government to companies for their production and operation activities. One ton of allowance equals one ton of CO_2 emissions.
(2) *Allowances to new entrants* represent the estimated annual CO_2 emissions allowances granted to the newly operated projects; such allowances are defined and issued by the development and reform commission in light of the CO_2 emissions evaluation report for these projects.
(3) *Reserve allowances* refer to the portion of allowances reserved by government for adjusting carbon prices so as to tackle the fluctuating carbon market and volatile economic situations; the quantity of such allowances is 5% of the combined allowances to the covered enterprises/institutions.
(4) Chinese Certified Emission Reduction (CCER) is generated from the greenhouse gas VER projects that are filed by the National Development and Reform Commission in light of the *Interim Measures for the Administration of Greenhouse Gas Voluntary Emission Reduction.*

Article 43 These measures shall come into effect as of March 1, 2014.

Carbon Emissions Allowance Administration and Implementation Rules in Guangdong Province

Chapter 1 General Provisions

Article 1 For the purpose of strengthening and standardizing the carbon emissions allowance administration in Guangdong Province, ensuring the authenticity, the *Carbon Emissions Allowance Administration and Implementation Rules in*

Guangdong Province are hereby formulated (hereinafter referred to as the "Implementation Rules") in light of the *Interim Measures for the Administration of Carbon Emissions Trading* (NDRC No. 17) and the *Interim Measures for Carbon Emissions Administration in Guangdong Province* (GD GOV No. 197) and based on the practical work for carrying out the pilot program for carbon emissions trading scheme.

Article 2 As regarding the covered enterprises/institutions (hereinafter referred to as "regulated companies"), new entrants (including expanded and rebuilt companies, similarly hereinafter), other qualified institutions and individuals within the jurisdiction of Guangdong Province, their carbon emissions allowance (hereinafter referred to as the "allowance"), clearing for compliance and transactions are subject to provisions of the Implementation Rules.

Article 3 The allowance administration of Guangdong Province shall abide by the following principles:

(1) Achieve emissions reduction and promote socioeconomic development. By means of allowance administration, Guangdong will be able to effectively control the greenhouse gas (GHG) emissions, fulfill the binding indicators for saving energy and cutting carbon emissions, and promote sustainable development of the entire province.
(2) Put efficiency in the first place and take account of fairness. In light of the provincial socioeconomic situations, Guangdong shall take full consideration of the benchmarking emissions of covered enterprises and their historical annual emissions, so as to ensure fair and reasonable allocation of allowances.
(3) Mainly rely on free allocation and gradually increase paid allowances. Guangdong shall combine both free and paid allocation of allowances, with free allocation playing a leading role at the preliminary stage of the ETS program, and then increase percentage of paid allowances step by step.
(4) Fair transaction and effective supervision. Fair transaction of the allowances shall be conducted on the carbon emissions trading platform (hereinafter referred to as the "trading platform") designated by the provincial government, the process of transaction shall be conscientiously supervised and managed.

Article 4 Being the competent authority responsible for the provincial allowance administration, the provincial development and reform commission takes charge of administration, supervision, and guidance of the ETS implementation, and works together with other competent departments to build an allowance allocation review board and a technical assessment panel for defining sectoral allowances, and then authorize the technical assessment panel or qualified social organizations/ institutions to carry out the related work.

All municipal development and reform commissions at and above the prefecture level shall collaborate with the provincial competent authority in reviewing the application from the new entrants with their administrative area for buying paid allowances, and urging the covered enterprises with their administrative area to carry out the allowances clearing and compliance.

Article 5 The provincial development and reform commission shall establish a carbon emissions allowance registration system in Guangdong Province (hereinafter referred to as the "registry") to take charge of unified electronic information management of allowance creation, allocation, alteration, clearing, and cancelation. The information recorded in the registry serves as final basis for defining ownership and obligation of allowances. The allowances will become valid as of the date of registration, since then the allowance owners are able to make earnings via trading, transfer or mortgage of the allowances or through other legitimate means.

Article 6 The registry shall create accounts of varied functions for the provincial competent authority, covered enterprises, new entrants, investment institutions and other market players. After opening the accounts in light of the requirements of the competent authority, the market players may start operations about allowances administration in the registry.

Chapter 2 Allowance Allocation

Article 7 The provincial development and reform commission, by taking a full account of the sectoral baseline emissions and emissions reduction potentials, and companies' historical emissions level, develops the allowance plan for Guangdong Province. Before published and enforced, the plan shall be first reviewed by the provincial allowance allocation review board and then ratified by the provincial people's government.

The allowance plan shall involve the name list of covered enterprises and new entrants, and amount of annual allowances; percentage of free and paid allowances; allocation methodologies, approaches, and processes; quantity of the paid allowances via bidding; and allocation platform and provisions.

Article 8 The technical assessment panel for defining sectoral allowances, which consists of the experts with industry associations and corporate representatives, takes charge of collecting, summarizing, and processing the opinions or suggestions reflected by the companies in the same industrial sector, evaluates the allowance calculation methodologies, emissions factor, benchmark value and annual depreciation factor in light of economic operation, and development characteristics of industries, and then submits the evaluation report to the provincial development and reform commission in a timely manner.

Article 9 In reference to production process, product characteristics, and data basis of an industrial sector, the provincial development and reform commission adopts the benchmarking and grandfathering methodologies to define the allowances to the covered enterprises and new entrants. The allowances granted to companies are a sum of the allowances for each production process (generating unit or product); their calculation formulas are shown as follows:

(1) Benchmarking:

Allowances allocated to covered enterprises = last year's actual production × correction factor for current year's production × benchmark value × current year's depreciation factor;
Allowances allocated to new entrants = designed production capacity × benchmark value.

(2) Grandfathering:

Allowances allocated to covered enterprises = historical average carbon emissions × current year's depreciation factor;
Allowances allocated to new entrants = projected annual comprehensive energy consumption × factor of converted carbon emissions.

The correction factor for sector production, benchmark value, and annual depreciation factor or calculation approaches shall be clarified in the annual allowance allocation plan.

Article 10 The provincial development and reform commission allocates free allowances to covered enterprises via the registry in light of the allowance allocation plan. Regarding the covered enterprises that receive benchmarking-based allowances, the provincial development and reform commission shall launch a province-wide emissions verification and then define the amount of allowances, and finally make up for the deficiency or take back excess allowances via the registry.

Article 11 The provincial development and reform commission organizes auctions for allocation of paid allowances in light of the allowance allocation plan. The covered enterprises may purchase the paid allowances voluntarily via the paid allowance bidding platform (hereinafter referred to as "bidding platform"). The new entrants, before transforming into covered enterprises, shall purchase sufficient paid allowances in light of the provisions in Chapter 4 in the Implementation Rules.

The revenue from paid allowance allocation shall be exclusively spent on the carbon emissions reduction and construction of relevant capacity.

Article 12 If the covered enterprises or new entrants disagree with the allocation results, they may apply for a reverification to the provincial development and reform commission which will send them a written reply within 20 days after reviewing the application.

If major changes happen to the production or operation of companies owing to changing economic situations or industrial development, the provincial development and reform commission shall authorize the technical assessment panel to evaluate their amount of allowances, and then draw up an overall program to adjust the allocation plan.

Chapter 3 Allowance Clearing and Compliance

Article 13 Before every June 20, the covered enterprises shall fulfill their compliance for turning over sufficient allowances via the registry based on last years' actual carbon emissions which are verified by the verification institution and ratified by the provincial development and reform commission. The allowances which are frozen for mortgage or financing may not be used for the clearing and compliance of the covered enterprises.

If the allowances are not enough to clear for compliance, the covered enterprises shall purchase certain allowances via the bidding platform in advance for making up for the deficiency. In case of any surplus allowances, after the covered enterprises have fulfilled their compliance, they shall be used for next year's compliance or trading.

Article 14 If the covered enterprises revoke their business license, halt production/operation and move out of the province, their allowances shall be handled as follows:

(1) The covered enterprises shall submit their carbon emissions information report and verification report to the municipal development and reform commissions at and above the prefecture level within one month before they halt production/operation or move out of the province, then the latter shall report such matter to the provincial development and reform commission.
(2) Before completion of the formalities for halting production/operation or moving out of the province, the covered enterprises shall clear the allowances based on the verified actual emissions of the current year, the free allowances for the companies, production lines, generating units or installations during abnormal production months (operation rate of the month is below 50%, similarly hereinafter) shall be handed over to the provincial development and reform commission to be canceled, while the remaining allowances may be used or traded by the companies.

Article 15 If the covered enterprises (production lines, generating units or installations) autonomously halt production for over 6 months in the current year, they shall hand over the free allowances for abnormal production months to the provincial development and reform commission for cancelation after confirmation.

Article 16 The covered enterprises are allowed to use the Chinese Certified Emission Reduction (CCER) or the carbon reductions of companies/institutions or individuals validated and issued by Guangdong Province to offset their actual carbon emissions. The CCER shall satisfy the relevant provisions in the *Interim Measures for Carbon Emissions Administration in Guangdong Province*, uniformly registered in light of the national relevant regulations and meeting the following conditions concurrently:

(1) Mainly come from the carbon dioxide (CO_2) and methane (CH_4) emissions reduction projects, i.e., these two gases shall account for over 50% of all GHG emissions cutbacks;
(2) The electricity generation, heat supply, and surplus energy (waste heat, water pressure, and waste gas) utilization projects that are free from consumption of hydropower or such fossil energy as coal, petroleum, and natural gas (excluding coal-bed methane); and
(3) Exclude the emissions reductions generated by the Clean Development Mechanism (CDM) projects before registered at the Executive Board of CDM of the United Nations.

Article 17 If the covered enterprises intend to use the CCER to offset their actual carbon emissions, they shall submit the offset application and relevant certifications to the provincial development and reform commission before every June 10 in light of the provisions of Article 16 in the Implementation Rules; the offset could not be carried out unless it is confirmed qualified.

The procedures for the offset with CCER or the carbon reductions of companies/institutions or individuals validated and issued by Guangdong Province are separately stipulated.

Article 18 If the covered enterprises fail to clear the allowances in full amount in violation of the provisions of the Implementation Rules, the provincial development and reform commission shall urge them to make correction; otherwise, the commission may impose a penalty upon such companies in light of the *Interim Measures for Carbon Emissions Administration in Guangdong Province*. And the violations of the companies will be input into the provincial financial credit system and social credit system according to relevant regulations, and make public via the official website of government and social media.

Chapter 4 Allowance Administration for New Entrants

Article 19 The companies that are newly operated with annual CO_2 emissions at 10,000 tons and above (hereinafter referred to as "new entrants") shall be included into allowance administration.

Article 20 The new entrants shall purchase paid allowances in line with the following procedures and requirements:

(1) The new entrants shall submit the application and relevant certification documents for buying paid allowances, the content of application involves but not limited to the following points:

i. Project profile;
ii. The energy-saving evaluation report and review opinions ratified or archived by the competent authority; and
iii. Projected annual CO_2 emissions of the new entrants.

(2) The application materials shall be first examined by the municipal development and reform commissions at and above the prefecture level according to the principle of territory, and then reported to the provincial development and reform commissions for validation. The provincial companies are able to report their application materials directly to the provincial development and reform commission for validation.
(3) Based on the verified quantity of paid allowances to the new entrants, the companies may purchase paid allowances in full amount via the bidding platform before acceptance check of the new projects.

Article 21 After the new entrants (production lines, generating units or installations) have fulfilled the acceptance check and undergone 12 months of operation, they shall be subject to verification of the verification institution designated by the provincial development and reform commission, those are fully qualified will officially join in the rank of covered enterprises.

If the new projects are owned by the existing covered enterprises, the latter shall revise their carbon emissions monitoring plan in a timely manner. After the new entrants (production lines, generating units or installations) have fulfilled the acceptance check and undergone 12 months of operation, they shall be incorporated into the existing covered enterprises.

Chapter 5 Allowance Transaction

Article 22 The trading principals joining in the allowance transaction are made up of covered enterprises, owners of new projects, qualified institutional, and individual investors. The transaction platform administration institution (hereinafter referred to as "transaction institution") shall specially build a professional transaction system that for allowance trading. The transaction system shall link with the allowance registry, enable data exchange, and make sure that the transaction information could be promptly inputted into in the registry.

Article 23 The institutional investors that join in the allowance transaction shall meet the following conditions concurrently:

(1) Being an independent legal entity;
(2) Being a well-organized institution with both internal control and risk management systems;
(3) With sound commercial credit; and
(4) Conform to the national and provincial relevant provisions.

Article 24 The individual investors that join in the allowance transaction shall meet the following conditions concurrently:

(1) Being a natural person at an age of 18 and above;
(2) Passing the transaction risk evaluation; and
(3) Conform to the national and provincial relevant provisions.

Article 25 The quantity of allowances held by each trading principal shall meet the following conditions:

(1) Before their annual actual carbon emissions are verified by the provincial development and reform commission, the covered enterprises may not transfer over 50% of their current year's free allowances in their registered account to the transaction account for trading. After clearing the allowances for compliance, the covered enterprises may transfer all of the surplus allowances to their transaction account for trading.
When the covered enterprises intend to deal with all of the allowances in their registered account owing to revocation of business license, halting production or removal out of the province, they shall go through the procedures as provided for in Article 14 in Chapter 3. The remaining allowances are at their own discretion.
(2) Before being incorporated into the covered enterprises, the owners of new projects may have unlimited quantity of allowances.
(3) The quantity of allowances held by institutional and individual investors may not exceed 3 Mt.

Article 26 All trading principals shall register at the registry and open an account at the transaction system at the same time. The provincial development and reform commission shall handle the application for opening an account made by the covered enterprises, the owners of new projects and overseas registered agencies, and authorize the transaction institution to handle the application for opening an account made by institutional and individual investors.

The transaction institution shall establish a transaction risk evaluation system, regularly report to the provincial development and reform commission about the transaction activities and approval results about the account opening applications made by institutional and individual investors.

Article 27 The provincial development and reform commission shall build allowance market regulation mechanism to safeguard market stability and supervise the business of the transaction institution. The transaction institution shall regularly evaluate and improve the transaction rules. After being stipulated, the transaction rules shall be first reported to the provincial development and reform commission and financial department for review and ratification before execution.

Article 28 The financial agencies are encouraged to explore the financial services like allowance mortgage finance, while such conduct shall be confirmed at the registry. The provincial development and reform commission shall entrust the transaction institution to do a good job in inquiring ownership and obligations of companies' allowances and mortgage registration.

Chapter 6 Supplementary Provisions

Article 29 The Implementation Rules are subject to interpretation of the provincial development and reform commission.

Article 30 Regarding the time limit as provided for in the Implementation Rules, if the date of expiration happens to be the national statutory holiday, then the first workday after the holiday is deemed as the date of expiration.

Article 31 The Implementation Rules come into effect as of March 1, 2015 and have a period of validity of five years. In case any relevant provision published before are inconsistent with the Implementation Rules, it is the latter that shall prevail.

Implementation Rules for Corporate Carbon Emissions Reporting and Verification in Guangdong Province

(The *Implementation Rules for Corporate Carbon Emissions Reporting and Verification in Guangdong Province* (GD DRC Climate Change Office [2015] No. 80) were issued by Guangdong Provincial Development and Reform Commission on February 16, 2015 and executed as of March 1, 2015.)

Chapter 1 General Provisions

Article 1 For the purpose of standardizing the corporate carbon emissions reporting and verification in Guangdong Province, ensuring the authenticity, accuracy, and reliability of corporate carbon emissions information, the *Implementation Rules for Corporate Carbon Emissions Reporting and Verification in Guangdong Province* (hereinafter referred to as the "Rules") are hereby formulated in light of the *Interim Measures for the Administration of Carbon Emissions Trading* and the *Interim Measures for Carbon Emissions Administration in Guangdong Province*.

Article 2 The Rules are applicable to the monitoring, reporting, and verification (MRV) of carbon emissions information of the covered enterprises/institutions within the administrative area of Guangdong Province.

Article 3 The provincial development and reform commission takes charge of overall coordination, supervision, and administration of the MRV of all of Guangdong's covered enterprises, and draws up carbon emissions reporting guidelines and verification specifications in collaboration of the provincial competent authorities. All municipal development and reform commissions at and above the prefecture level shall take charge of the MRV of the covered enterprises within their jurisdiction.

Article 4 The MRV regime of Guangdong Province deals with the covered enterprises/institutions and reporting companies based on their category. The specific name list of these companies/institutions is defined by the provincial development and reform commission and other relevant authorities through deliberation.

By taking account of the carbon emissions administration and progress of carbon trading, such industrial sectors as electricity, cement, iron and steel, petrochemical, ceramics, nonferrous, textile, chemical, and paper-making, as well as the public buildings and the companies/institutions that engage in transportation, are included into the MRV regime in batches. Any company/institution that intends to participate in the carbon emissions administration voluntarily shall report to the provincial development and reform commission for approval, and then join in the MRV regime in light of the same requirements for the covered enterprises/institutions.

Article 5 If the annual actual carbon emissions of the covered enterprises are verified to be lower than the prescribed amount for 3 years in a row, they will be degraded to reporting companies after confirmed by provincial development and reform commission. If the annual actual carbon emissions of the reporting companies are lower than the prescribed amount for 3 years in a row, they will be removed out of the carbon emissions administration after confirmed by provincial development and reform commission.

If the annual actual carbon emissions of the covered enterprises/institutions and reporting companies are lower than the prescribed amount, which is resulting from major changes on their production or operation activities, they shall be degraded to reporting companies or removed out of the carbon emissions administration after confirmed by provincial development and reform commission. Such major changes refer to altered production capacity (e.g., dismantled production line), production scope or product mix, or serious accident, technical revamp, and plant relocation which halts production for 2 years or more.

Article 6 The covered enterprises/institutions and reporting companies shall establish and improve the carbon emissions administration system, and perform monitoring and reporting obligations as required. The covered enterprises/institutions shall collaborate with the verification institutions to carry out the verification activities.

Article 7 The provincial development and reform commission shall build the Carbon Emissions Information Reporting and Verification System (hereinafter referred to as the "Information System") to take charge of information management of the MRV of companies/institutions.

Chapter 2 Carbon Emissions Monitoring and Reporting

Article 8 The carbon emissions monitoring plan is an important basis for the covered enterprises/institutions and reporting companies to report their emissions information, and for the verification institutions to carry out the verification activities. The covered enterprises/institutions and reporting companies shall produce an emissions monitoring plan and submit to the provincial development and reform commission together with their emissions report which is delivered for the first time.

Article 9 The carbon emissions monitoring plan, which is produced on basis of the *Carbon Dioxide Emissions Reporting Guidelines for Companies in Guangdong Province*, is made up of the following key points:

(1) Basic information about the emitting sources, including process flow diagram and inventory of main equipment;
(2) Available monitoring devices, approaches, and measurement frequency;
(3) Patterns for data recording, keeping statistics, processing, summarizing and storing;
(4) Quality control and guarantee measures; and
(5) Other relevant information.

In case any change occurs to the above contents, the covered enterprises/institutions and reporting companies shall modify their carbon emissions monitoring plan and submit the revised plan to the provincial development and reform commission within 3 months after the change takes place.

Article 10 The covered enterprises/institutions and reporting companies shall produce an annual emissions monitoring plan and deliver the plan via the aforesaid Information System by March 15 every year. The reporting companies shall also deliver an official monitoring plan in written plan to the municipal development and reform commission at and above the prefecture level. The covered enterprises/institutions shall, before May 5 each year, deliver the verified annual emissions report and verification report, both in written form, to the municipal development and reform commission at and above the prefecture level which shall summarize the reports and then deliver them to the provincial development and reform commission before May 10 each year.

Article 11 The carbon emissions information report, which is produced on basis of the *Carbon Dioxide Emissions Reporting Guidelines for Companies in Guangdong Province*, is made up of the following key points:

(1) The basic information about companies' production and operation;
(2) The information about the energy, materials, and production process associated with companies' carbon emissions;
(3) The parameters, methodologies, and results of carbon emissions metering;
(4) Data quality control and guarantee;
(5) Submit the entrustment agreement if a service agency is entrusted to produce the carbon emissions information report; and
(6) Other matters that need to be explained.

Article 12 The covered enterprises/institutions shall cooperate with the verification institution to carry out the relevant work, provide necessary documents and materials, then modify the carbon emissions information report in reference to the opinions presented by the verification institution, and finally resubmit the improved report.

Chapter 3 Carbon Emissions Verification

Article 13 The verification institution shall, by following the verification specifications and instructions of Guangdong Province, carry out the verification activities by means of document review and on-site check, and submit the verification report via the Information System by April 30 each year, and produce a written verification report to the covered enterprises/institutions.

Article 14 According to the regulations on government procurement, the provincial development and reform commission may entrust a verification institution to verify (including reverification and spot check) the emissions information report submitted by the covered enterprises/institutions, and bear the corresponding expenses.

Article 15 If any change happens to the verification institution, including the changes on the legal representative, premises, business scope, verification scope, and verifiers, it shall report such matter to the provincial development and reform commission within ten workdays after the change takes place. If the verification institution is unable to satisfy the verification requirements, provincial development and reform commission shall confirm such matter and no longer invite this verification institution.

Article 16 Before carrying out the verification activities, the verification institution shall report the name list of the concerned verifiers to the provincial development and reform commission, and organize the concerned verifiers to take part in the verification trainings sponsored by the commission.

The verification trainings held by the provincial development and reform commission are free from charge.

Article 17 In light of the verification specifications of Guangdong Province, the verification institution shall produce a verification report and affix their signatures and seals on the report for confirmation. The verification report consists of the following key points:

(1) Basic information about the verification institution;
(2) Information about the personnel composition of the verification team;
(3) Basic information about the verified companies/institutions;
(4) Record of the verification process;
(5) Annual carbon emissions amount of companies/institutions and methodologies for emissions calculation;
(6) Verification conclusion;
(7) The content of the emissions information report that needs to be clarified or revised;

(8) Impartiality statement;
(9) Other matters that need to be explained.

Article 18 The provincial development and reform commission shall hold a review of the emissions information report produced by the covered enterprises/institutions and the verification report. If there is no dissent opinion about these reports, the commission shall confirm the annual emissions amount of the covered enterprises/institutions and deliver the feedback to them before May 20 each year; otherwise, the commission shall organize a reverification and feed back the annual emissions amount to the covered enterprises/institutions before June 5 each year.

Article 19 Upon receipt of the feedback from the provincial development and reform commission, if the covered enterprises/institutions hold different views toward the defined annual emissions amount, they are entitled to apply for a reverification to the commission within seven workdays upon receipt of the feedback.

Chapter 4 Supervision and Administration

Article 20 The covered enterprises/institutions take responsibility for the authenticity of their emissions monitoring plan, reported data and information. If any regulated or reporting company/institution fails to receive the mandatory emissions verification, they shall be fined in light of the *Interim Measures for Carbon Emissions Administration in Guangdong Province*.

The provincial development and reform commission shall strengthen supervision and administration of the carbon emissions monitoring plan produced by the covered enterprises/institutions, and present opinions for modification if the plan is found inconsistent with the actual carbon emissions.

If any regulated or reporting company/institution fails to report their carbon emissions or not collaborate with the verification institution, the provincial development and reform commission shall entrust the verification institution to calculate their annual carbon emissions, which shall be deemed as basis for the concerned company/institution to perform their obligations in allowance settlement.

Article 21 The provincial development and reform commission shall exercise dynamic administration of the verification institution, organize assessment and appraisal of the verification report, and establish a "blacklist" system for the non-compliant verification institutions.

Article 22 When the provincial development and reform commission or all municipal development and reform commissions at and above the prefecture level are performing their obligation of supervising the verification report, the concerned companies/institutions shall collaborate during the supervision process by providing relevant files and materials, instead of rejection, obstruction or concealment.

Article 23 The verification institution and its staff members shall carry out the verification activity according to law and in an independent and impartial manner; take responsibility for the standardization, authenticity, and accuracy of the verification reform; and perform the obligation of confidentiality, bear legal liability, and accept public supervision. If the verification institution or its staff members commit one of the following actions, the provincial development and reform commission shall make public their violation and impose a fine in light of the *Interim Measures for Carbon Emissions Administration in Guangdong Province*; those whose offenses are serious enough to constitute a crime shall be prosecuted for criminal responsibility according to law.

(1) Produce a false and inconsistent verification report;
(2) Serious mistakes are found in the verification report or the verified emissions are greatly different from actual emissions.
(3) Use or publish the commercial secrets and carbon emissions information of the verified companies/institutions without authorization;
(4) Any other acts of violations.

Article 24 The provincial development and reform commission shall create a credit file for the covered enterprises/institutions, reporting companies and verification institutions, and incorporate their credit information into the Social Credit System and Financial Credit System in a timely manner.

Article 25 If the employees with the development and reform commissions at all levels and competent authorities commit any of the misconduct like abuse of power, neglect of duty, practicing frauds for personal gains or disclose the confidential information about the covered enterprises/institutions shall be imposed an administrative sanction or disciplinary sanction. Those whose offenses are serious enough to constitute a crime shall be prosecuted for criminal responsibility according to law.

Article 26 The provincial development and reform commission, covered enterprises/institutions, reporting companies and verification institutions shall make archive for the carbon emissions report, verification report and relevant files. Such materials shall be kept by the provincial development and reform commission for 30 years, while such materials shall be kept by covered enterprises/institutions, reporting companies, and verification institutions for 15 years.

Chapter 5 Supplementary Provisions

Article 27 The Rules are due to be explained by the provincial development and reform commission.

Article 28 If the last calendar day of the deadline as prescribed in the Rules happens to be a statutory holiday, then the last workday ahead of the holiday is deemed as the date of expiration.

Article 29 The Rules shall come into effect as of March 1, 2015 and have a period of validity of 5 years. In case the provisions in the Rules are inconsistent with the provisions that are published before, it is the Rules that shall prevail.

Carbon Emissions Trading Rules of China (Guangzhou) Emissions Exchange

Chapter 1 General Provisions

Article 1 For the purpose of standardizing the carbon emissions trading activities, maintaining market order, and safeguarding the legitimate rights and interests of market players, the *Carbon Emissions Trading Rules of China (Guangzhou) Emissions Exchange* are hereby formulated in light of the *Interim Measures for Carbon Emissions Administration in Guangdong Province*.

Article 2 The Rules are applicable to China (Guangzhou) Emissions Exchange (hereinafter referred to as "CGEE") that hosts and organizes the carbon emissions trading activities of Guangdong Province.

Article 3 The parties that engage in CGEE-based carbon trading shall observe the relevant laws and regulations, and the rules and provisions of CGEE, and follow the principles of openness, fairness, impartiality, voluntariness, equality, honesty, and credibility.

Chapter 2 Trading Market

Section 1 Venue for Transaction

Article 4 CGEE is obliged to provide a transaction venue, relevant facilities, and services for the carbon trading parties.

Article 4 A transaction venue and associated facilities shall consist of a trading hall, data center, information disclosure system, settlement system, and a supporting system relevant to carbon trading.

Section 1 Participants in Trading Activities

Article 6 Trading participants, which refer to all parties that purchase or sell the carbon emissions allowances in CGEE, are made up of the following parties:

(1) The covered enterprises/institutions and new entrants that are involved in the carbon emissions scheme (ETS) in Guangdong Province; and
(2) Other investment institutions, organizations, and individuals that conform to relevant regulations.

Article 7 CGEE has established a membership administration system. The trading parties shall become a CGEE member or entrust a CGEE member to take part in the allowance trading. CGEE shall report the name list of the members to the provincial development and reform commission in a regular manner.

The membership administration for emissions allowance trading is separately provided for by CGEE.

Article 8 The trading parties are entitled to the following rights:

(1) Join in the emissions allowance trading and associated activities;
(2) Attend the relevant trainings organized by CGEE;
(3) Use the relevant equipment and facilities provided by CGEE;
(4) Receive the information and services that concern about the emissions allowance trading that are provided by CGEE;
(5) Supervise the work of CGEE and present suggestions or opinions;
(6) Other legitimate rights that are granted by law.

Article 9 The trading parties shall perform the following obligations:

(1) Abide by relevant laws and regulations, the Rules, and any other administrative systems that are established by CGEE;
(2) Properly preserve the transaction account and password, bear the legal liability for the trading directive issued from the transaction account and the transaction outcome, and bear all liability for the consequences generated by the use of the transaction account;
(3) Learn about the information, announcements, and systems that are disclosed by CGEE, and bear the losses arising from underperformed obligation of reasonable concern;
(4) Bear the corresponding risks and legal liabilities for the contract that stipulated by trading parties, and strictly perform the contract and join in emissions allowance trading in a fairness, impartial, and open manner;
(5) Take care of the CGEE-based facilities, safeguard its reputation, and pay for various expenses based on the contract;
(6) Notify CGEE of the major event that may impact the allowances trading;
(7) Ensure the authenticity, integrity, and validity of the submitted materials, and undertake corresponding legal liabilities therefor;
(8) Perform other obligations according to law.

Section 3 Objects and Specifications of Transaction

Article 10 The objects of transaction of CGEE mainly consist of the following products:

(1) The carbon emissions allowances of Guangdong Province (GDEA);
(2) Other tradable products that are approved by the provincial development and reform commission.

Article 11 The emissions allowance trading in CGEE is measured in the following basic units:

(1) Trading unit: per ton of carbon dioxide emissions (tCO_2);
(2) Price quotation unit: yuan/t (rounded up to percentile);
(3) Minimum trading volume: one ton;
(4) Minimum price fluctuation unit: 0.01 yuan/t.

Section 4 Time of Transaction

Article 12 The transactions in CGEE are held from 9:30 to 11:30 a.m. and from 13:30 to 15:30 p.m. from Monday to Friday (subject to the time of the transaction server). No transaction is held on the national statutory holidays or any closure day as announced by CGEE.

If it is required for market development, CGEE is allowed to adjust time of transaction and make an announcement about this matter.

Article 13 No transaction is postponed even if a market closure takes place within the trading hours for some reason.

Chapter 3 Trading of Carbon Emissions Allowance

Article 14 Carbon emissions allowance is traded via listed bidding, click choosing, one-way bidding, negotiated transfer, and any other approaches ratified by the provincial development and reform commission.

The viable trading approach is decided by CGEE in light of specific situations and announced to the public.

Article 15 All trading parties shall open an allowance transaction account via the registry of CGEE, and also open a transaction account and settlement account in their real name as required.

Article 16 Before sending out the declaration orders to the transaction system, all trading parties shall make sure that the allowances or funds in their transaction account are enough to meet the transaction conditions.

Article 17 Upon completion of the trading activity, the transaction system will automatically generate an electronic transaction voucher which is of the corresponding legal effect.

Section 1 Listed Bidding

Article 18 In listed bidding, the trading parties shall send out the declaration orders via the transaction system which will then rank and disclose the declaration orders, and finally arrange one-time and one-way matching for the declaration orders within the time span as prescribed by the transaction system.

Article 19 The transaction time span for listed bidding is for sending out declaration orders and paired trading.

During the declaration time span, the trading parties shall send out the declaration orders to the transaction system. The declaration orders are made up of code of transaction object, quantity, price, and trade direction. Upon completion of the declaration, the corresponding transaction object and funds will be frozen.

During the paired trading time span, the transaction system shall, by following the principle of "price priority and time priority", carry out the paired trading, and disclose the final result. Any untraded declaration order will enter into the next transaction time span.

Article 20 The transferor and transferee may cancel their declaration orders within the declaration time span, then the frozen transaction object or fund will be unfreezed automatically; the completed declaration orders may not be canceled; as for the partially completed declaration orders, only the uncompleted portion may be canceled. During the paired trading time span, neither declaration nor withdrawal is allowed. The untraded declaration orders will be canceled automatically after the transaction hours of listed bidding.

Section 2 Click Choosing

Article 21 In click choosing, the trading parties shall propose listed declarations for sales or purchase to define the quantity and price of transaction object, the intentional transferor and transferee shall view the real-time listed orders, click on the intentional order, and submit the declarations for sales or purchase.

Article 22 The trading parties shall submit the listed declaration orders to the transaction system for attracting intentional transferor or transferee. The listed orders shall include the code of transaction object, quantity, unit price, and trading direction. Upon completion of the declaration, the corresponding transaction object or fund will be frozen and enter into the queue of listed orders.

The intentional transferor or transferee shall view the real-time listed declaration orders, click on the intentional orders, submit the declaration, and complete the trading. By following the principle of "price priority and time priority", the intentional transferor is only allowed to click on the minimum priced sales order, while the transferee shall click on the maximum price purchase order. The order is closed based on its declared price and quantity.

Article 23 All of the untraded listed declarations may be canceled any time. For the partially traded listed declarations, only the untraded portion may be canceled and available to be listed again.

Section 3 One-Way Bidding

Article 24 In one-way bidding, transferor shall submit the listed orders to the transaction system, define quantity, and reserve price of the transaction object, and the intentional transferee shall conduct online bidding on their own and finish the transaction with the prescribed time limit.

Article 25 The transferor shall submit the listed orders for one-way trading to the transaction system for attracting intentional transferee. The listed orders shall include the code of transaction object, quantity, reserve price, and time limit for the bidding. Upon completion of the declaration, the corresponding transaction object will be frozen and enter into the bidding process at the time set by the transaction system.

The bidding process is made up of free bidding and time-bound bidding. During free bidding, the intentional transferor is able to give full quotations for the transaction object; when the quotation price is higher than the existing highest quotation price (opening price is allowed for the first quotation), it is deemed as valid quotation price. After completion of the free bidding process, the transaction system will postpone the preset limit cycle automatically, and enter into the time-bound bidding process which shall be within the trading hours as announced by the CGEE. Upon completion of the time-bound bidding, the latest valid quotation price will become the strike price.

Section 4 Negotiated Transfer

Article 26 Negotiated transfer refers to a transaction pattern where the two trading parties complete the transaction by reaching consensus through negotiations. When the negotiated transfer is practiced, the quantity of single transaction shall reach 100,000 tons or above.

Article 27 In case of negotiated transfer, the quotation price may not be 130% higher or 70% lower than the closing price of the previous trading day.

Article 28 The trading parties submit the listed orders based on negotiated transfer to the transaction system. In addition to the code of transaction object, quantity, price, and trade direction, the information about the intentional transferee shall be included. The transaction is completed upon confirmation by the intentional transferee in the transaction system and approval of CGEE.

Article 29 The trading price based on negotiated transfer is not included in the real-time quotations of CGEE; yet the trading volume thereof is reckoned into CGEE's gross trading volume of the allowances on that day.

Chapter 4 Fund Supervision, Clearing, and Allowance Delivery

Article 30 CGEE has established a third-party custody system for clearing the funds for allowance transaction.

Neither institution nor individual may embezzle, occupy, or borrow the transaction clearing funds that are deposited in bank in the name of CGEE, or use the funds for providing guarantee for others without CGEE's approval.

Article 31 CGEE shall deal with the transaction clearing after completion of the transactions on the same day. The allowances shall be transferred from the transaction account of transferor to that of transferee, while the funds shall be transferred by the clearing bank from the transaction account of transferee to that of transferor.

The transferor is able to transfer the clearing funds from its transaction account into its fund account on the following trading day.

Article 32 The allowance delivery is uniformly organized by CGEE. The provincial development and reform commission shall, in light of relevant regulations and the clearing results of CGEE, ratify the transfer of the ownership of allowances via the allowance registry.

Article 33 CGEE arranges fund clearing and allowance delivery from 15:30 to 17:00 p.m. on each trading day. In case the time for fund clearing and allowance delivery is altered under a special circumstance, it shall be separately announced by CGEE.

The system for allowance clearing is separately stipulated by CGEE.

Chapter 5 Other Transaction Matters

Section 1 Opening Price, Closing Price, and Increase/Decrease Rate

Article 34 The opening price at CGEE is the closing price of listed bidding or click choosing of the previous trading day, while the closing price is weighted average price of aggregate trading prices of the listed bidding or click choosing on the same day. In case of no closing price or transaction on that day, the closing price of the previous trading day is deemed as the valid closing price.

Article 35 The closing price of listed bidding or click choosing shall be ±10% of the opening price. In case the one-way bidding is practiced, the reserve price shall be ±10% of the opening price.

Section 2 Suspension, Resumption, and Termination of Transactions

Article 36 In case one of the following circumstances takes place, CGEE shall suspend the transaction of certain allowances or the entire transaction system. In case any transaction is made and leads to a serious consequence, CGEE may adopt appropriate remedial measure and deem it invalid.

(1) Partial or whole transaction is undone owing to technical breakdown or unlawful intrusion of transaction system, contingency, or matters of force majeure;
(2) Any transaction that is suspected of violating law or relevant regulations, trading volume exceeding the prescribed scope, high-frequency transaction, or other abnormal situations;
(3) Any other circumstance that interrupts normal operation of transactions or the circumstance where CGEE deems necessary to suspend transactions.

When CGEE decides to suspend transactions, it shall report such decision to the provincial development and reform commission and make announcement about such matter.

Article 37 CGEE may decide to resume transactions after dispelling the interrupting circumstances. The specific timing and approach for suspending or resuming the transactions are decided by CGEE in light of the specific situations.

Article 38 During the period where the transactions are suspended, the transaction system no longer accepts the declarations about the suspended transactions or all of the declarations. Upon resumption of the transactions, the directives made before the suspension shall enter the normal trading process.

Article 39 In case the trading parties or allowances no longer have the trading qualifications or conditions in light of the relevant laws, regulations, rules or policies, CGEE shall terminate the relevant transaction activities and make announcement about such matter.

Article 40 CGEE is exempted from bearing the liability for the losses arising from transaction suspension, resumption or termination.

Chapter 6 Transaction Information

Article 41 CGEE shall publish the real-time quotations and public information about the allowance transactions on each trading day.

Article 42 CGEE shall regularly produce the statistical statement and analysis report about the allowance transactions and release them on time.

Article 43 CGEE, trading parties, and clearing bank may not disclose the commercial secrets that they have obtained when dealing with the allowance transactions.

CGEE is allowed to provide relevant information to the competent authority or other relevant institutions in light of relevant regulations and carry out the regulations on commercial secrets.

Article 44 CGEE is allowed to adjust the means for publishing information and its content in line with the need for market development.

The system for administration of allowance transaction information is separately stipulated by CGEE.

Chapter 7 Supervision and Administration

Article 45 CGEE has established the system for limiting the holdings of allowances. The allowances held by each trading party may not exceed the prescribed limit.

Article 46 CGEE has established the system for control of clearing risks, and exercises administration of members' clearing funds based on separate accounts.

Article 47 CGEE shall conduct supervision and inspection of the trading parties, clearing bank, and other participants in allowance transaction in light the provisions in these Rules and other relevant regulations, and regularly report such matters to the provincial development and reform commission.

Article 48 If CGEE has discovered that the trading parties are involved in insider dealing, market manipulation, abnormal transactions, and other violations by means of supervision, complaints reporting or notification of competent authority, it shall order the concerned party to make correction; in light of the seriousness of the case, CGEE may impose such penalties as conversation reminder, written warning, public criticism, restriction of transaction, suspension or termination of transactions, or revocation of their qualifications for other businesses or membership.

CGEE shall preserve a poor credit record for all trading parties, record their violations during transactions, and make public such discreditable conducts.

Chapter 8 Resolution of Transaction Disputes

Article 49 In case any dispute arises from the allowance transaction, the concerned trading parties may resolve such matter on their own through negotiations, apply for arbitration by an arbitration institution according to law or file a lawsuit to the local people's court.

Chapter 9 Transaction Expenses

Article 50 When the trading parties join in the allowance transactions held by CGEE, or any institution or individual uses the information provided by CGEE, they shall pay for the transaction service, information use, and other necessary fees to CGEE.

Article 51 The amount of allowance transaction formality fee is decided in line with the relevant regulations of the provincial pricing authority, while other charging items and standards are separately decided by CGEE.

Chapter 10 Supplementary Provisions

Article 52 These Rules are stipulated and explained by CGEE. Based on the provisions of these Rules, CGEE is able to draw up implementation provisions or measures.

Article 53 These Rules shall come into effect as of the date of issuance.

China (Guangzhou) Emissions Exchange
July 30, 2015

Trading Rules for Chinese Certified Emission Reduction at China (Guangzhou) Emissions Exchange

Chapter 1 General Provisions

Article 1 For the purpose of standardizing the transactions of Chinese Certified Emission Reduction (hereinafter referred to as "CCER"), maintaining market order and safeguarding the legitimate rights and interests of market players, the *Trading Rules for Chinese Certified Emission Reduction at China (Guangzhou) Emissions Exchange* are hereby formulated in light of the *Interim Measures for the Administration of Greenhouse Gas Voluntary Emission Reduction* and the *Interim Measures for Carbon Emissions Administration in Guangdong Province*.

Article 2 The Rules are applicable to China (Guangzhou) Emissions Exchange (hereinafter referred to as "CGEE") that hosts and organizes the CCER trading activities.

The parties that engage in CGEE-based CCER trading shall observe the relevant laws and regulations, and the rules and provisions of CGEE, and follow the principles of openness, fairness, impartiality, voluntariness, equality, honesty, and credibility.

Article 3 The CCER that is referred to in these Rules represents the Chinese Certified Emission Reduction that is ratified by the National Development and Reform Commission (hereinafter referred to as the "NDRC").

Chapter 2 Trading Market

Section 1 Venue for Transaction

Article 4 CGEE is obliged to provide a transaction venue, relevant facilities, and services for the carbon trading parties.

Article 5 A transaction venue and associated facilities shall consist of a trading hall, data center, information disclosure system, settlement system, and a supporting system relevant to carbon trading.

Section 2 Participants in Trading Activities

Article 6 Trading participants, which refer to all parties that purchase or sell the CCER in CGEE, are made up of the following parties:

(1) Owners of CCER;
(2) The covered enterprises/institutions and new entrants that are involved in the pilot program for carbon emissions scheme (ETS) in Guangdong Province; and
(3) Other investment institutions, organizations, and individuals that conform to relevant regulations.

Article 7 CGEE has established a membership administration system. The trading parties shall become a CGEE member or entrust a CGEE member to take part in the CCER trading.

The membership administration for CCER trading is separately provided for by CGEE.

Article 8 The trading parties are entitled to the following rights:

(1) Join in the CCER trading and associated activities;
(2) Attend the relevant trainings organized by CGEE;
(3) Use the relevant equipment and facilities provided by CGEE;
(4) Receive the information and services that concern about the CCER trading that are provided by CGEE;
(5) Supervise the work of CGEE and present suggestions or opinions;
(6) Other legitimate rights that are granted by law.

Article 9 The trading parties shall perform the following obligations:

(1) Abide by relevant laws and regulations, the Rules, and any other administrative systems that are established by CGEE;
(2) Properly preserve the transaction account and password, bear the legal liability for the trading directive issued from the transaction account and the transaction outcome, and bear all liability for the consequences generated by the use of the transaction account;
(3) Learn about the information, announcements, and systems that are disclosed by CGEE, and bear the losses arising from underperformed obligation of reasonable concern;
(4) Bear the corresponding risks and legal liabilities for the contract that stipulated by trading parties, and strictly perform the contract and join in emissions allowance trading in a fairness, impartial, and open manner;
(5) Take care of the CGEE-based facilities, safeguard its reputation, and pay for various expenses based on the contract;
(6) Notify CGEE of the major event that may impact the CCER trading;
(7) Ensure the authenticity, integrity, and validity of the submitted materials, and undertake corresponding legal liabilities therefor;
(8) Perform other obligations according to law.

Section 3 Object and Specifications of Transaction

Article 10 In these Rules, the object of transaction of CGEE refers to CCER.

Article 11 The CCER trading in CGEE is measured in the following basic units:

(1) Trading unit: per ton of carbon dioxide emissions (tCO_2);
(2) Price quotation unit: yuan/t (rounded up to percentile);
(3) Minimum trading volume: one ton;
(4) Minimum price fluctuation unit: 0.01 yuan/t.

Section 4 Time of Transaction

Article 12 The transactions in CGEE are held from 9:30 to 11:30 a.m. and from 13:30 to 15:30 p.m. from Monday to Friday (subject to the time of the transaction server). No transaction is held on the national statutory holidays or any closure day as announced by CGEE.

If it is required for market development, CGEE is allowed to adjust time of transaction and make an announcement about this matter.

Article 13 No transaction is postponed even if a market closure takes place within the trading hours for some reason.

Chapter 3 CCER Trading

Article 14 CCER is traded via listed bidding, click choosing, one-way bidding, negotiated transfer, and any other approaches ratified by the competent authority.

The viable trading approach is decided by CGEE in light of specific situations and announced to the public.

Article 15 All trading parties shall open a CCER transaction account via the registry of CGEE, and also open a transaction account and settlement account in their real name as required.

Article 16 Before sending out the declaration orders to the transaction system, all trading parties shall make sure that the CCER or funds in their transaction account are enough to meet the transaction conditions.

Article 17 Upon completion of the trading activity, the transaction system will automatically generate an electronic transaction voucher which is of the corresponding legal effect.

Section 1 Listed Bidding

Article 18 In listed bidding, the trading parties shall send out the declaration orders via the transaction system which will then rank and disclose the declaration orders, and finally arrange one-time and one-way matching for the declaration orders within the time span as prescribed by the transaction system.

Article 19 The transaction time span for listed bidding is for sending out declaration orders and paired trading.

During the declaration time span, the trading parties shall send out the declaration orders to the transaction system. The declaration orders are made up of code of transaction object, quantity, price, and trade direction. Upon completion of the declaration, the corresponding transaction object and funds will be frozen.

During the paired trading time span, the transaction system shall, by following the principle of "price priority and time priority", carry out the paired trading, and disclose the final result. Any untraded declaration order will enter into the next transaction time span.

Article 20 The transferor and transferee may cancel their declaration orders within the declaration time span, then the frozen transaction object or fund will be unfreezed automatically; the completed declaration orders may not be canceled; as for the partially completed declaration orders, only the uncompleted portion may be canceled. During the paired trading time span, neither declaration nor withdrawal is allowed. The untraded declaration orders will be canceled automatically after the transaction hours of listed bidding.

Section 2 Click Choosing

Article 21 In click choosing, the trading parties shall propose listed declarations for sales or purchase to define the quantity and price of transaction object, the intentional transferor and transferee shall view the real-time listed orders, click on the intentional order, and submit the declarations for sales or purchase.

Article 22 The trading parties shall submit the listed declaration orders to the transaction system for attracting intentional transferor or transferee. The listed orders shall include the code of transaction object, quantity, unit price, and trading direction. Upon completion of the declaration, the corresponding transaction object or fund will be frozen and enter into the queue of listed orders.

The intentional transferor or transferee shall view the real-time listed declaration orders, click on the intentional orders, submit the declaration and complete the trading. By following the principle of "price priority and time priority", the intentional transferor is only allowed to click on the minimum priced sales order, while the transferee shall click on the maximum price purchase order. The order is closed based on its declared price and quantity.

Article 23 All of the untraded listed declarations may be canceled any time. For the partially traded listed declarations, only the untraded portion may be canceled and available to be listed again.

Section 3 One-Way Bidding

Article 24 In one-way bidding, transferor shall submit the listed orders to the transaction system, define quantity, and reserve price of the transaction object, the intentional transferee shall conduct online bidding on their own and finish the transaction with the prescribed time limit.

Article 25 The transferor shall submit the listed orders for one-way trading to the transaction system for attracting intentional transferee. The listed orders shall include the code of transaction object, quantity, reserve price, and time limit for the bidding. Upon completion of the declaration, the corresponding transaction object will be frozen and enter into the bidding process at the time set by the transaction system.

The bidding process is made up of free bidding and time-bound bidding. During free bidding, the intentional transferor is able to give full quotations for the transaction object; when the quotation price is higher than the existing highest quotation price (opening price is allowed for the first quotation), it is deemed as valid quotation price. After completion of the free bidding process, the transaction system will postpone the preset limit cycle automatically, and enter into the time-bound bidding process which shall be within the trading hours as announced by the CGEE. Upon completion of the time-bound bidding, the latest valid quotation price will become the strike price.

Section 4 Negotiated Transfer

Article 26 Negotiated transfer refers to a transaction pattern where the two trading parties complete the transaction by reaching consensus through negotiations.

Article 27 The trading parties submit the listed orders based on negotiated transfer to the transaction system. In addition to the code of transaction object, quantity, price, and trade direction, the information about the intentional transferee shall be included. The transaction is completed upon confirmation by the intentional transferee in the transaction system and approval of CGEE.

Article 29 The trading price based on negotiated transfer is not included in the real-time quotations of CGEE; yet the trading volume thereof is reckoned into CGEE's gross trading volume of the CCER on that day.

Chapter 4 Fund Supervision, Clearing, and CCER Delivery

Article 29 CGEE has established a third-party custody system for clearing the funds for CCER transaction.

Neither institution nor individual may embezzle, occupy or borrow the transaction clearing funds that are deposited in bank in the name of CGEE, or use the funds for providing guarantee for others without CGEE's approval.

Article 30 CGEE shall deal with the transaction clearing after completion of the transactions on the same day. The allowances shall be transferred from the transaction account of transferor to that of transferee, while the funds shall be transferred by the clearing bank from the transaction account of transferee to that of transferor.

The transferor is able to transfer the clearing funds from its transaction account into its fund account on the following trading day.

Article 31 The CCER delivery is uniformly organized by CGEE. The provincial development and reform commission shall, in light of relevant regulations and the clearing results of CGEE, ratify the transfer of the ownership of CCER via the CCER registry.

Article 32 CGEE arranges fund clearing and CCER delivery from 15:30 to 17:00 p.m. on each trading day. In case the time for fund clearing and allowance delivery is altered under a special circumstance, it shall be separately announced by CGEE.

The system for CCER clearing is separately stipulated by CGEE.

Chapter 5 Other Transaction Matters

Section 1 Opening Price, Closing Price, and Increase/Decrease Rate

Article 33 The opening price at CGEE is the closing price of listed bidding or click choosing of the previous trading day, while the closing price is weighted average price of aggregate trading prices of the listed bidding or click choosing on the same day. In case of no closing price or transaction on that day, the closing price of the previous trading day is deemed as the valid closing price.

Article 34 The closing price of listed bidding or click choosing shall be ±10% of the opening price. In case the one-way bidding is practiced, the reserve price shall be ±10% of the opening price.

Section 2 Suspension, Resumption, and Termination of Transactions

Article 35 In case one of the following circumstances takes place, CGEE shall suspend the transaction of certain CCER or the entire transaction system. In case any transaction is made and leads to a serious consequence, CGEE may adopt appropriate remedial measure and deem it invalid.

(1) Partial or whole transaction is undone owing to technical breakdown or unlawful intrusion of transaction system, contingency, or matters of force majeure;
(2) Any transaction that is suspected of violating law or relevant regulations, trading volume exceeding the prescribed scope, high-frequency transaction, or other abnormal situations;
(3) Any other circumstance that interrupts normal operation of transactions or the circumstance where CGEE deems necessary to suspend transactions.

When CGEE decides to suspend transactions, it shall report such decision to the provincial development and reform commission and make announcement about such matter.

Article 36 CGEE may decide to resume transactions after dispelling the interrupting circumstances. The specific timing and approach for suspending or resuming the transactions are decided by CGEE in light of the specific situations.

Article 37 During the period where the transactions are suspended, the transaction system no longer accepts the declarations about the suspended transactions or all of the declarations. Upon resumption of the transactions, the directives made before the suspension shall enter the normal trading process.

Article 38 In case the trading parties or allowances no longer have the trading qualifications or conditions in light of the relevant laws, regulations, rules, or policies, CGEE shall terminate the relevant transaction activities and make announcement about such matter.

Article 39 CGEE is exempted from bearing the liability for the losses arising from transaction suspension, resumption or termination.

Chapter 6 Transaction Information

Article 40 CGEE shall publish the real-time quotations and public information about the CCER transactions on each trading day.

Article 41 CGEE shall regularly produce the statistical statement and analysis report about the CCER transactions and release them on time.

Article 42 CGEE, trading parties, and clearing bank may not disclose the commercial secrets that they have obtained when dealing with the CCER transactions.

CGEE is allowed to provide relevant information to the competent authority or other relevant institutions in light of relevant regulations and carry out the regulations on commercial secrets.

Article 43 CGEE is allowed to adjust the means for publishing information and its content in line with the need for market development.

The system for administration of CCER transaction information is separately stipulated by CGEE.

Chapter 7 Supervision and Administration

Article 44 CGEE has established the system for control of clearing risks, and exercises administration of members' clearing funds based on separate accounts.

Article 45 CGEE shall conduct supervision and inspection of the trading parties, clearing bank, and other participants in CCER transaction in light the provisions in these Rules and other relevant regulations, and regularly report such matters to the provincial development and reform commission.

Article 46 If CGEE has discovered that the trading parties are involved in insider dealing, market manipulation, abnormal transactions, and other violations by means of supervision, complaints reporting or notification of competent authority, it shall order the concerned party to make correction; in light of the seriousness of the case, CGEE may impose such penalties as conversation reminder, written warning, public criticism, restriction of transaction, suspension or termination of transactions, or revocation of their qualifications for other businesses or membership.

CGEE shall preserve a poor credit record for all trading parties, record their violations during transactions, and make public such discreditable conducts.

Chapter 8 Resolution of Transaction Disputes

Article 47 In case any dispute arises from the CCER transaction, the concerned trading parties may resolve such matter on their own through negotiations, apply for arbitration by an arbitration institution according to law or file a lawsuit to the local people's court.

Chapter 9 Transaction Expenses

Article 48 When the trading parties join in the CCER transactions held by CGEE, or any institution or individual uses the information provided by CGEE, they shall pay for the transaction service, information use, and other necessary fees to CGEE.

Article 49 The amount of CCER transaction formality fee is decided in line with the relevant regulations of the provincial pricing authority, while other charging items and standards are separately decided by CGEE.

Chapter 10 Supplementary Provisions

Article 50 These Rules are stipulated and explained by CGEE. Based on the provisions of these Rules, CGEE is able to draw up implementation provisions or measures.

Article 51 These Rules shall come into effect as of the date of issuance.

China (Guangzhou) Emissions Exchange
January 7, 2016

Printed in the United States
By Bookmasters